Quinazolinone and Quinazoline Derivatives

Edited by Ali Gamal Al-kaf

Published in London, United Kingdom

IntechOpen

Supporting open minds since 2005

Quinazolinone and Quinazoline Derivatives
http://dx.doi.org/10.5772/intechopen.85315
Edited by Ali Gamal Al-kaf

Contributors
Heba Hashem, Samir Youssef Abbas, Cornel Chirita, Carmen Limban, Diana Camelia Nuţă, Emil Ştefănescu, Cristina Elena Zbârcea, Simona Negreş, Cristina Daniela Marineci, Mihaela Bratu, Alexandru Vasile Missir, Ileana Cornelia Chiriţă, Oana Cristina Seremet, Satyendra Mishra, Sainath Zangade, Ali Gamal Al-kaf, Awwad Radwan, Fars Kaed Alanazi

First published in London, United Kingdom, 2020 by IntechOpen
IntechOpen is the global imprint of INTECHOPEN LIMITED, registered in England and Wales, registration number: 11086078, 7th floor, 10 Lower Thames Street, London, EC3R 6AF, United Kingdom
Printed in Croatia

British Library Cataloguing-in-Publication Data
A catalogue record for this book is available from the British Library

Additional hard and PDF copies can be obtained from orders@intechopen.com

Quinazolinone and Quinazoline Derivatives
Edited by Ali Gamal Al-kaf
p. cm.
Print ISBN 978-1-83880-139-7
Online ISBN 978-1-83880-140-3
eBook (PDF) ISBN 978-1-83880-053-6

For EU product safety concerns:
IN TECH d.o.o., Prolaz Marije Krucifikse Kozulić 3, 51000 Rijeka, Croatia, info@intechopen.com or visit our website at intechopen.com.

Meet the editor

Professor Doctor Ali Gamal Al-kaf has a PhD in Pharmaceutical Sciences from the Chief Council for Accreditation and Quality Assurance, Russia (2006). Previously he has been Dean of Faculty of Pharmacy at Sana'a University, Professor of the Medicinal Chemistry Department, member of the Yemeni Medical Council, and a member of many associations and international groups. He is executive editor of the *Universal Journal of Pharmaceutical Research*, and editor and associate editor of a number of international journals. His interests are the synthesis and biological activity of 4-oxopyrimidine and quinazolinone-4 derivatives, structural biology and bioinformatics in drug design, study of Yemeni medicinal plants, development and validation of spectrophotometric and HPLC methods for different drugs, and antibiotic and antimicrobial resistance in Yemen. .He is the author of more than 70 publications, four patents, and 11 books.

Contents

Preface

This book consists of many chapters on different topics in the field of the latest research in quinazolinone and quinazoline derivatives.

The aim is to focus on all the methods for the synthesis and different interesting biological activities of quinazolinone and quinazoline derivatives.

The major objective of this book is to present the information in a lucid, condensed, and cohesive form, and to cater especially to the needs of readers in medicine and pharmacy.

The book has seven chapters from authors all over the world and covers the following topics:

• Introductory Chapter: The Newest Research in Quinazolinone and Quinazoline Derivatives

• Biological Activity of Quinazolinones

• Synthesis of Quinazoline andQuinazolinone Derivatives

• 4(3H)-Quinazolinone Derivatives: Syntheses, Physical Properties, Chemical Reaction, and Biological Properties

• Quinazolinone and Quinazoline Derivatives: Synthesis and Biological Application

• Synthesis and Pharmacological Research Regarding New Compounds with Quinazolin-4-One Structure

• Quinazoline Based Synthesis of some Heterocyclic Schiff Bases

Many thanks go to all the authors for their valuable, interesting, and important research into quinazolinone and quinazoline derivatives.

The cooperation of the publisher IntechOpen is very much appreciated in publishing this book. The assistance received from Ms. Jasna Bozic, Author Service Manager, cannot be ignored. Any suggestions, comments, and criticisms on the subject matter will be gratefully acknowledged to improve future editions. My hope is that this work will prove to be a benefit to students and teachers of pharmacy and science, and to medical scientists.

Dr. Ali Gamal Al-kaf
Professor,
Medicinal Chemistry Department,
Chief Council for Accreditation and Quality Assurance,
Member of Higher Yemeni Medical Council,
Previous Dean of Faculty of Pharmacy-Sana'a University,
Yemen

Section 1

The Newest Research in Quinazolinone and Quinazoline Derivatives

Chapter 1

Introductory Chapter: The Newest Research in Quinazolinone and Quinazoline Derivatives

Ali Gamal Al-kaf

1. Introduction

Target searching for new high effective medicinal preparation is considered one of the actual problems of the modern public health. From medicinal preparations are the natural and synthetic origins of quinazolinone-4 derivatives. Quinazolinone derivatives are reported to be physiologically and pharmacologically active [1].

They also exhibit a wide range of activities such as anticonvulsant, anti-inflammatory, antifungal, antimalarial, and sedative. Some of these compounds are identified as drugs used as diuretics, vasodilators, and antihypertensive agents.

Moreover, sulfonamide derivatives have been widely used as bacteriostatic agents. Prompted by the abovementioned facts and in conjunction with our ongoing program on the utility of readily obtainable starting material for the synthesis of heterocyclic systems of biological interest, we have decided to synthesize a series of quinazolinone derivatives having sulfonamide moiety with potentially wide spectrum of biological responses [2, 3]. Information about the biological properties of derivatives of quinazolinone before the end of 60-years ago was more fragmentary.

At the end of 40 years, an alkaloidal compound febrifugin (dichroin) 3-[β-keto-γ(3-oxipiperidine-2) propyl] quinazolone-4 was isolated from Chinese plant Chang-Shan (Dichroa febrifuga Laur.) and has antimalarial activity, 100 times greater than quinine. This compound has not found clinical application, since it is 300 times toxic than quinine alkaloids [4, 9, 10, 21] (**Figure 1**).

However, work with febrifugin encouraged further searching for new biologically active compounds among the derivatives of quinazolinones-4 [4].

Already, on early stages of the studies, chemotherapeutic, anti-inflammatory activity, and hypnotic and hypotensive actions were revealed [5].

Sixty to seventy years ago, more than 300 patents appeared in different countries, denoted synthesis and study of biological activity of this type of compounds. More than 20 medicinal preparations (analgesic, sedative, diuretic, hypotensive,

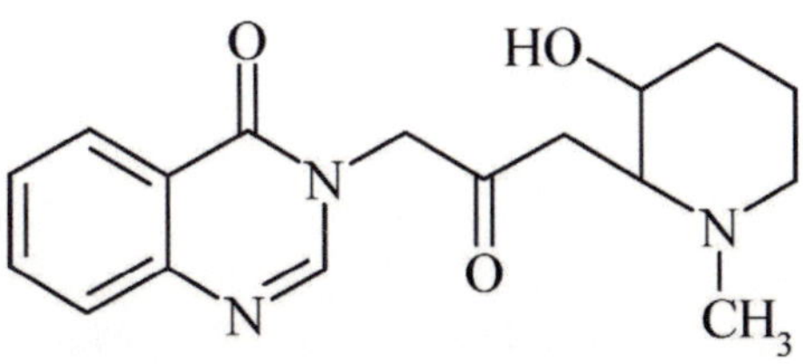

Figure 1.
The structure of the compound febrifugin (dichroin).

cholegogue, bronchodilator, tranquilizer, chemotherapeutic action) of quinazoline derivatives are introduced in medicinal practice [5–7].

Among various quinazolinone derivatives, big attention was attracted to the substituted-(3H) quinazolinones-4, which show the significant hypnotic activity [4].

2-Methyl-3-(o-tolyl) quinazolinone-4, named methaqualone, is broadly used in medicinal practice [5, 8] as sedative and hypnotic. Aside from hypnotic action, methaqualone possesses the anticonvulsant and anti-cough properties and intensifies the action of barbiturates, analgesics, and neuroleptics [7].

The high activity and small toxicity of methaqualone were a motivation for study of the large number of its analogues [9] (**Figure 2**).

Similar analogues of methaqualone such as—a light hypnotic ethaqualone, as well as mecloqualone which is used in medicine as hypnotic and sedative [4, 10].

2-Methyl-3-(o-cyan phenyl) quinazolone-4 is characterized by the highest hypnotic, sedative, and anticonvulsant activity [11]. Strong psychotropic action is showed in triflourmethyl analogue of methaqualone [12].

The series analogues of methaqualone with similar pharmacological properties with ortho-substiuted have an alkyls or other functional groups and in the other positions phenyl residues [4, 13].

Among compounds of this group, the most popular compounds that have been used in practical medicine are 2-methyl-3-(o-methyl-p-chlorphenyl)-5-chlorquinazolone-4 (SL-I64) as tranquilizer and 2-methyl-3-(o-methoxy-p-nitrophenyl) quinazolone-4 (nitromethaqualone) as hypnotic [14]. Quinazolinones-4 render also hypothermic and spasmolytic action [5].

At study of more than 40 derivatives of quinazolines, some qualitative correlations between structure and convulsive activity were revealed [15, 16]. Transition from 3-aryl- to 3-alkylquinazolones-4 saves the sedative, hypnotic, and anticonvulsant properties of compounds [16]. The last type of activity is also characterized for 2-methyl-3-piperazino-alkylquinazolones-4.

Variation of the substituted in position 3 may affect upon biological properties of substances in these compounds. So, 3-cyanalkyl-quinazolones-4, for example, are diuretics, but corresponding to aminoketones and aminohydroxy, derivatives show analgesic, spasmolytic, and anti-inflammatory activities [4].

Derivatives of (1H)-quinazolinones-4 are less studied. For them, the most characterized properties are analgesic, anti-cough, and anti-inflammatory activity [10, 17]. 1,2,3,4-Tetrahydroquinazolones-4 were studied greatly in details.

Some compounds of this line, unsubstituted in position 1, show the antibacterial and cytostatic activities. 1-Alkyl or aryl derivatives are characterized as anti-inflammatory, bronchodilator, and sedative action; 1-dialkylaminoalkyl- and 1-acyl, including substituted aminoacyl-1-tetrahydroquinazolones-4, have manifested themselves as analgesics and tranquilizers, myorelexants, and diuretics.

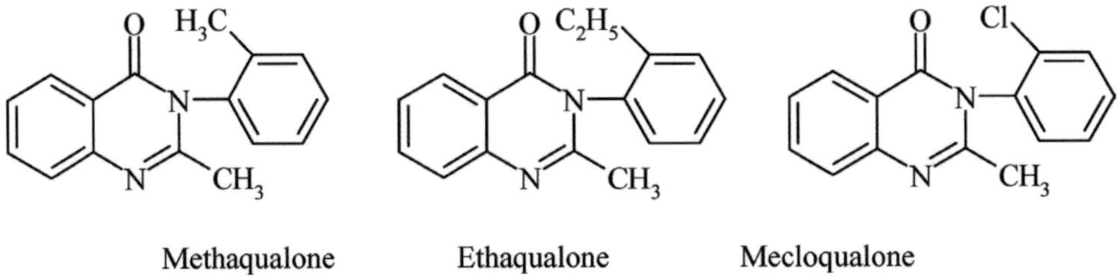

Figure 2.
The structures of methaqualone and its analogues.

Significantly, their parts described antihistamine, cholagogue, and anti-inflammatory properties [10].

N-Heteril derivatives of quinazolinone-4, containing of fragments of antipyrin and thiazole in the position 3, have anti-hypoxic activity [18].

Anagrelide is a quinazoline compound, possessing anti-thrombocyte aggregation activity [19]. Quinazoline-derived doxazosin and terazosin (α_1-adrenoreceptor antagonists) show the highest activity in the cancer of the prostate gland [19, 20].

Mokvizon [21] has been found to be used in practical medicine as cholagogue and hypocholesterolemic remedy (**Figure 3**).

Among 3-alkyliden-1,2.3,4-tetrahydroquinazolones-4, glycozine was used as preparation of hypnotic action.

Introduction of chlorine in position seven and sulfamide group in position six of quinazolines-4 brings the high active diuretics and saluretics of the chlorothiazide type. Diuretic action of the substances increases when they are turning from quinazolones-4 to tetrahydroquinazolones [9].

Compounds, for example, of this type are metolazone, SR 720-22, zaroxoline, quinethazone, hydromox, and others. They are introduced into medicinal practice as diuretics and saluretics [4].

Pharmacological screening has allowed revealing the sedative activity for some derivatives of quinazolinone-4:

4-(2-Phenyl-4-oxoquinazolil) –butyric acid increases the duration of nembutalic sleep in comparison with control on 106% [22, 23] and influences upon parameters of hemodynamic system of awake rats [24–26].

Derivatives of quinazoline-4(3H)-one show antimicrobial, antifungal, antimalarial, antituberculosic, antihypertensive activity, as well as anticonvulsant and sedative actions [27–30].

The pharmacological studies found that the most activity, that these compounds have 2′, 4′-dimethoxyphenyl radical at position 2 of quinazolone cycle.

So, the most anti-inflammatory activity is characterized by 1-acetyl-2-(2′, 4′-dimethoxyphenyl)-3-[(4-methoxyphenyl) or (4-chlorphenyl)]-1,2,3, 4-tetrahydroquinazolinones-4.

High anticonvulsant activity was shown in 1-adamantilacetyl-2-(2′,4′-dimethoxyphenyl)-3-benzyl-1,2,3,4-tetrahydroquinazolinone-4 [27].

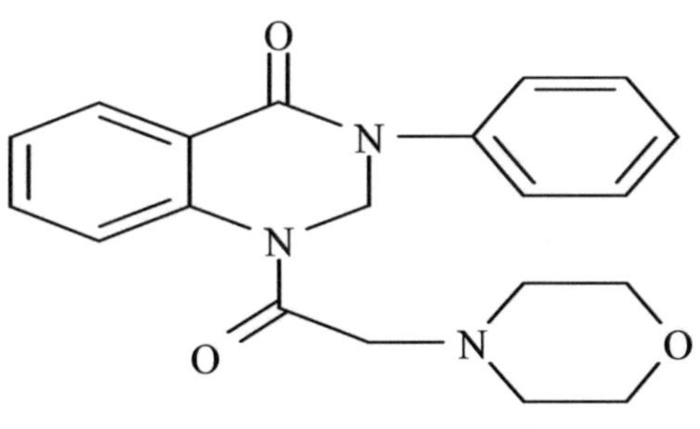

Mokvizon

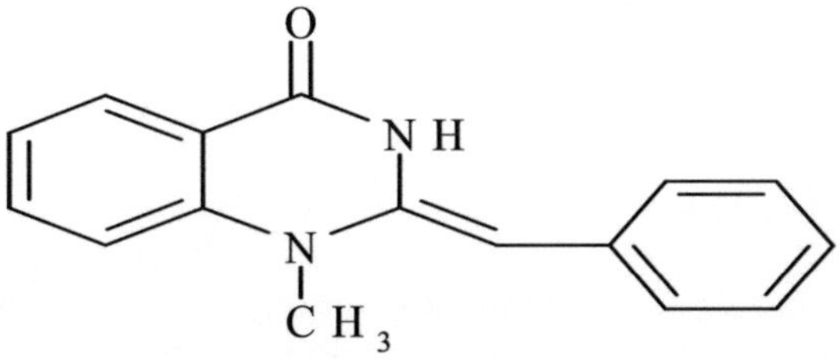

Glycozine

Figure 3.
The structures of mokvizon and glycozine.

2-Styril-4-aminoquinazolinones are characterized by strong activity in respect of gram-positive bacteria, *Mycobacterium tuberculosis*, pathogenic proterozoic infection, pathogenic viruses, and fungi [30].

Pharmacological screening of some derivatives of quinazolinone-4 has allowed revealing for analeptic, anti-inflammatory, and anti-hypoxic activity and slightly toxic:

4-Amino-4′-[2-phenyl-4-oxoquinazoline-3]-diphenylsulfone (QPhD) and 4,4'-bis-[2-phenyl-4-oxoquinazoline-3]-diphenylsulfone (BisQPhD) have analeptic, anti-inflammatory, and anti-hypoxic activities and are slightly toxic [21, 31].

In one study 4-amino-4′-[2-phenyl-4-oxoquinazoline-3]-diphenylsulfone (QPhD) shows strong immunotropic activity surpassing the comparing drug methyluracil [32].

In one study 4-(4-Oxo-2-phenyl-4H-quinazolin-3-yl)-N-pyrimidin-2-yl-benzenesulfonamide (Compound A) and N-(4,6-Dimethyl-pyrimidin-2-yl)-4-(4-oxo-2-phenyl-4H-quinazolin-3-yl)-benzenesulfonamide (Compound B) show strong hepatoprotective and antioxidant activity surpassing the comparing drug Liv-52 [33].

Accessibility of quinazoline derivatives and versatility of their biological activity attract well-earned attention to similar class of substances. The similarity is not only physicochemical properties of pyrimidine and quinazolinone, but also spectrum of biological activity opens the new possibilities for searching for high active compounds in these class substances [10, 34]. The extensive array of information, denoted synthesis, and study of the biological activity of derivatives of quinazolinon-4 is indicative of exclusive possibility of the synthesis of new biologically active compounds. Synthesis of 4-(4-oxo-2-phenyl-4H-quinazolin-3-yl)-N-pyrimidin-2-yl-benzenesulfonamide (Compound A) and N-(4,6-dimethyl-pyrimidin-2-yl)-4-(4-oxo-2-phenyl-4H-quinazolin-3-yl)-benzenesulfonamide (Compound B) showed high antimalarial activity compared with chloroquin [35].

The synthesis of 4-(4-oxo-2-phenyl-4H-quinazolin-3-yl) benzene sulfonamide (Compound A) and N-acetyl-4-(4-oxo-2-phenyl-4H-quinazolin-3-yl)-benzenesulfonamide (Compound C) having analgesic and anti-inflammatory activities but Compound C showed a significant activity than Compound A; also the Compound A has a pharmacological activity more significant than standard drug diclofenac. Further studies are important to ensure their analgesic and anti-inflammatory mechanisms.

The docking study explains the inhibitory effect of the A and C compounds against the in vivo anti-inflammatory and analgesic activity. The compounds A and C have similar effect as diclofenac and indomethacin reference drugs in the in vivo anti-inflammatory and analgesic activity [36].

Synthesis of 4-(4-oxo-2-phenyl-4H-quinazolin-3-yl)-N-pyrimidin-2-yl-benzene sulfonamide (Compound B) and N-(4,6-dimethyl-pyrimidin-2-yl)-4-(4-oxo-2-phenyl-4H-quinazolin-3-yl)-benzene sulfonamide (Compound D) gave a Pa < 0.5 as nootropic by the PASS program, but experiments on animals confirmed the anti-hypoxic activity of the compounds, which means they might occur as new chemical entities. Compound D has shown the strongest anti-hypoxic activity. A docking study of the synthesized derivatives with GluA3 confirmed the result in vitro and revealed that Compound D is an anti-hypoxic agent [37].

Our aim is to focus on all the methods for synthesis and different interesting biological activities of quinazolinone and quinazoline derivatives.

Our major objective of this project is to give the information in a lucid, condensed, and cohesive form and to specially cater the needs of readers in medicine and pharmacy.

Author details

Ali Gamal Al-kaf
Medicinal Chemistry Department, Faculty of Pharmacy, Sana'a University, Sana'a, Yemen

*Address all correspondence to: alialkaf21@gmail.com

References

[1] Pojarsky AF. Theoretical principles of heterocyclic chemistry. Chemistry. 1985:276

[2] Ram VJ, Tripathi BK, Srivastava AK. Bioorganic & Medicinal Chemistry. 2003;**11**:2439

[3] Dandia A, Singh R, Sarawgi P. Synthesis of some Annelated Quinazoline Derivatives. The Journal of the Chinese Chemical Society. Vol. 445. China, Taipei: Wiley-VCH & The Chemical Society. 2005. p. 307

[4] Vizgunova OL, Kojevnicov UV, Obvinseva LM, Zalesov VS. Study amongst quinazolinone-4. XVII. Synthesis and biological activity of 1, 2- disubstituted of quinazolinone-4. Chemical-The Pharmaceutical Journal. 1985;**33**(11):1047-1049

[5] Morgaluk VP et al. Synthesis and cardiotonic activity derivatives of 6, 7-dimethoxyquinazoline. Chemical-Pharmaceutical Journal. 1991;**25**(1):28-32

[6] Kang-Chien L, Ji-Wang C. Synthesis and antihypertensive activity of some fused quinazoline derivatives. Heterocycles. l984;**21**:503

[7] Mashkovskiy MD. The Medicinal Remedies. Vol. 2. Kharkov: Torsing; 1997. p. 592

[8] Wilson and Gisvold's. Textbook of Organic Medicinal and Pharmaceutical Chemistry. 10th ed. New York: Lippincott Raven; 1998. p. 446

[9] Chernobrovin NE, Kojevnicov UV, Bobrovskaya OV, Siropchtov BYA. Synthesis and biological activity of 1-acetyl-2,3-diaryl-1,2,3,4-tetrahydroquinazolinones-4. Chemical The Pharmaceutical Journal. 1999;**30**(11):40-44

[10] Yahontov LN, Liberman SS, Jihareva GP, Kuzmin KK. Biological activity of quinazoline derivatives. Chemical-Pharmaceutical Journal. 1977;**11**:14-26

[11] Ribuhin UE. Salts of 4 (5H)-oxazolonium, 4-oxo-1,3-oxazinonium and its hetero-analogues. [D.Sc. Thesis, doctor of chemical sciences]. Rostov-Na Danu; 1991. pp. 184-187

[12] Kojevnicov UV, Smirnov NN, Zalesov VS, Gradel EE. Study amongst 4[3h] quinazolone. X. Synthesis of biological activity of perchlorates 1-ethyl-2-methyl-3-aryl-4[3h] quinazolonium. Chemical-Pharmaceutical Journal. 1981;**12**(6):55-59

[13] Yahontov LN et al. Synthesis and study of biological activity of substituted 4-amino-2-styrylquinazolines. Chemical-Pharmaceutical Journal. 1975;**9**(11):12-18

[14] Martindale: The Extra Pharmacopoeia. 30th ed. Vol. 3750. London: Pharmaceutical Press; 1993

[15] Kojevnicov UV, Smirnov NN, Zalesov VS, Gradel EE. Study amongst 4[3h] quinazolone. X. Synthesis and biological activity of perchlorates 1-ethyl-2-methyl-3-aryl-4[3h] quinazolonium. Chemical-Pharmaceutical Journal. 1981;**12**(6):55-59

[16] Bakibaev AA et al. Synthetic anticonvulsants, anticonvulsants and inductors of monoxygenase system of the liver on the bases of amides and urea. V. Synthesis and biological activity of new derivatives of ortho—substituted benzylhydrilurea—Potential syntones for quinazolines and benzazepines.

Chemical-Pharmaceutical Journal. 1991;**28**(7):11-12

[17] Nogrady T. Medicinal Chemistry. 2nd ed. Vol. 514. New York: Oxford; 1988

[18] Oganesyan ET et al. The prediction and targeted synthesis of N-heterocyclic derivatives of quinazolinone-4. In: Modern High Technology Techniques. Vol. 6. 2004. pp. 109-110

[19] Robak T, Trelinski J, Holub A. Anagrelide-new anti-platelet drug. Acta Haematologica Polonica. 1994;**25**(4):309-315

[20] Drug Information for the Health Care Professional. 22nd ed. Vol. 1. New York; 2002. p. 3291

[21] Castle RN. Condensed Pyridazines Including Chinnolines and Phthalazines. Vol. 98. New York: Wiley-Interscience; 1973

[22] Kodonidi EP et al. Pharmacological screening of derivatives quinazolinone-4. Development, research and marketing of new pharmaceutical production: Scient. Pyatigorsk: Works collect. Of PSPA; 2003. 58. pp. 365-367

[23] Kodonidi EP et al. Prediction and synthesis of derivatives of quinazolinone-4 with fragments of carbonic acids at N3 position of the heterocycle person and a medicine: Person and a medicine: Thesis of the 10th Russian international congress report. 2-6 April 2003. 2003. pp. 642

[24] Kodonidi EP et al. Effect of new derivative of quinazolinone on the parameters of hemodynamic system in a wake rats. Development, Research and Marketing of New Pharmaceutical Production: Scient. Works Collect. Of PSPA. Pyatigorsk, 2003. 58. pp. 364-365

[25] Kodonidi EP et al. Effect of new derivative of 4-oxopyrimidine and quinazolinone-4 on some parameters system and cerebral hemodynamic. Development, research and marketing of new pharmaceutical production: Scient. Pyatigorsk: Works collect of PSPA; 2004. 59. pp. 246-247

[26] Magonov MM. Synthesis and biological activity of derivatives of 4-oxopyrimidine and quinazolinone-4 with fragments of carbonic acids in nucleotide position [PhD thesis]. Pyatigorsk: Pharmaceutical Dep. Academy of Pharm; 2002. pp. 117

[27] Chernobrovin NE et al. Synthesis and biological activity of 1-acetyl-2, 3-diaryl-1,2,3,4-tetrahydroquinazolinones-4. Chemical-Pharmaceutical Journal. 1991;**25**(5):37

[28] Somer F et al. Original 2-alkylamino-6-halogenoquinazolin-4 (3H)-ones and K (ATP) channel activity. Journal of Medicinal Chemistry. 2001;**44**(16):2575-2585

[29] Kreutzberger A, Balbach S. 2-Guanidino-4 (3H)-quinazolinones with chemotherapeutic activities. Die Pharmazie. 1993;**48**(1):17-20

[30] Agrawal VK et al. QSAR studies on antimalarial 2,4-diamino-6-quinazoline sulfonamide. Department of Chemistry. 2001;**48**(1):17-26

[31] Ji-Wang C, Chia-Yang S, Guan-Yu L. Studies on quinazolinones. 3: Novel and efficient route to the synthesis of conformationally rigid analogues of ketanserin and SGB-1534 as antihypertensive agents. Biomedical Chemistry Letters. 1991;**1**:571

[32] Jihareva GP et al. Synthesis and chemotherapeutic study of 2-styryl-4-(δ-diethyl amino-α-methyl butyl amino) quinazolines. Chemical-Pharmaceutical Journal. 1982;**16**(2):183-188

[33] Yassina SH, Al-kafa AG, Kotbb M, Abdullaha JH, Yahyaa TA. Antimicrobial, anti-oxidant and hepatoprotective activities of quinazolinone-4 derivatives on paracetamol induced hepatotoxicity in Guinea pigs. Journal of Natural and Applied Sciences. 2014;**18**:1

[34] Hemminki K. Ring-opened 7-methylguanine nucleotides are resistant to nuclease P1 digestion and good substrates to polynucleotide kinase. Carcinogenesis. 1989;**10**(9):1761-1763

[35] Al-kaf A et al. Synthesis and antimalarial of new quinazolinone-4 derivatives. Research in Pharmacy and Health Sciences. 2016;**2**(4):227-233

[36] Al-kaf A, Jihane A, Zahr BF, Rachid EJ, M'Hammed A, Azeddine I, et al. Synthesis, analgesic, antinflammatory and molecular docking studies of new quinazolin-4-one derivatives. Journal Marine Chemistry Heterocyclic. 2017;**16**(1)

[37] Akachar J, Abudunia A-M, El Jaoudi R, Bennani FE, Ansar M, Ibrahimi A, et al. Synthesis, antihypoxic activity and molecular docking studies of new quinazolinone-4 derivatives. Journal of Science and Technology. 2018;**23**:5-16

Chapter 2

Biological Activity of Quinazolinones

Awwad A. Radwan and Fars K. Alanazi

Abstract

The chemical structure of quinazolinones includes benzene ring fused with 2-pyrimidinone (1), 4-pyrimidinone (2) or 2,4-pyrimidinedione (3) ring, and are named as quinazolin-2(1H)-one, quinazolin-4(3H)-one or quinazolin-2,4(1H, 3H)-one, respectively. The chemical structure of quinazolinones constitutes a crucial scaffold of natural and synthetic compounds with various therapeutic and biological activities. Quinazolinones are first synthesized by Stefan Niementowski (1866–1925) and named after Niementowski quinazolinone synthesis. Quinazolinones have strongly attracted the interest of medicinal chemist as they constitute a large class of compounds that exhibited broad spectrum of biological activities including antimicrobial, antimalarial, anticonvulsant, anticancer, anti-leishmanial, anti-inflammatory, etc. This chapter provides a brief overview on the recent advances on chemical and pharmacological aspects of quinazolinone derivatives published in the last decade.

Keywords: quinazolinones, antimicrobial, antimalarial, anticancer

1. Introduction

Heterocyclic compounds are organic cyclic compounds having at least one atom other than carbon in their ring structures. Quinazolinones are formed by fusion of benzene ring with 2-pyrimidinone (1), 4-pyrimidinone (2) or 2,4-pyrimidinedione (3) ring, and are named as quinazolin-2(*1H*)-one, quinazolin-4(*3H*)-one or quinazolin-2,4(*1H, 3H*)-one, respectively (**Figure 1**).

Quinazolinones are pharmacophoric scaffold ubiquitous in various biologically active natural products, synthetic compounds, pharmaceutical drugs, agrochemicals and veterinary products [1]. The chemical structure of quinazolinones constitute a crucial scaffold of compounds with various therapeutic and biological activities such as antimalarial [2], antimicrobial [3, 4], antitubercular [5], anticonvulsant [6], anticancer [7], antihypertensive [8], anti-diabetic [9], anti-inflammatory [10], anti-cholinesterase [11], cellular phosphorylation inhibition [12], dihydrofolate reductase inhibition [13], kinase inhibitory activities [14], inhibitors of tubuline polymerization [15], diuretic [16], antipsychotic [17], dopamine agonists [18] and anti-HIV [19]. Quinazolinones (**Figure 1**) is a core scaffold for the structure of more than 200 naturally occurring alkaloids isolated from different plant families and from various microorganisms such as *Bacillus cereus, Peganum nigellastrum, Dichroa febrifuga and Bouchardatia neurococca* [20]. Depending on the position of the keto group, they can be classified into three types. Among the 2(1H)-quinazolinones (**1**), 4(3H)-quinazolinones

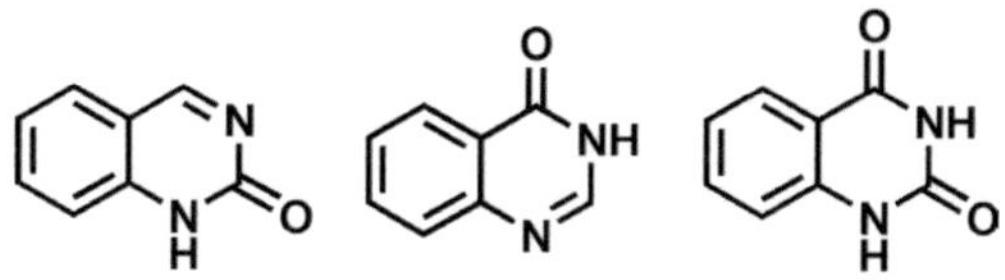

(1) (2) (3)

Figure 1.
Quinazolinone and quinazolinedione structures.

and 2,4(1H,3H)-quinazolinediones (**2**), 4(3H)-quinazolinones (**3**) are most prevalent and significant in medicinal chemistry possessing a multitude of pharmacological actions [21]. Quinazolinones are generally classified into four classes, 2-substituted quinazolinones, 3-substituted quinazolinones, 2,3-disubstituted quinazolinones and quinazolinone derivatives including fused quinazolinones [22].

2. Synthetic methods of quinazolinones

The number of synthetic methods of quinazolinone cores has intensely increased from year to year. These advancements in methods of synthesis lead to the access to new and effective quinazolinone compounds with augmented structural diversity starting from affordable and easily accessible substrates. In this chapter, we depict different methods of synthesis of quinazolinone derivatives from cheap and readily available starting precursors.

2.1 Synthesis of quinazolinone compounds from 2-aminobenzoic acid

Quinazolin-4(3*H*)-one (4) was synthesized by the reaction of formamide with 2-aminobenzoic acid at 125–130°C and cyclization of 2-aminobenzoic acid takes place as described in **Figure 2** [23]. Synthetic works started from esterification of 2-aminobenzoic acid and subsequently followed by reaction with isocyanates afforded 1,3-disubstituted quinazol-2,4(*1H, 3H*)-diones (5) [24] (**Figure 2**). 2-mercapto-3-substituted quinazolin-4(3*H*)-one derivatives (6) (**Figure 2**) have been synthesized through the interaction between 2-aminobenzoic acid and corresponding isothiocyanate reagent.

In 1960, Ried et al. reported [25–27] the reaction of imidates and 2-aminobenzoic acid in methanol at 80°C to afford the desired quinazolinones (7) in good yields (**Figure 2**).

A recently reported route, to the synthesis of 2-substituted quinazolin-4(3H)-ones (7) under microwave conditions was reported by Rad-Moghadam and Mohseni [28]. This approach involves the condensation of 2-aminobenzoic acid, orthoesters and ammonium acetate which afford the 2-substituted-4(3H)-quinazolinone (7) **Figure 2**.

A solvent-free approach was reported by Li et al. [29] for the synthesis of 2,3-disubstituted-4(*3H*)-quinazolinones (8). The approach involves the interaction between 2-aminobenzoic acid, acyl chlorides and aromatic/aliphatic amines in the presence of SO_3H-functionalized Brønsted acid ionic liquids as a catalyst under microwave irradiation (**Figure 2**). Langer and Döring [30] reported the reaction of 2-aminobenzoic acids with oxalic acid bis(imidoyl) chlorides to prepare quinazolinones (9) **Figure 2**.

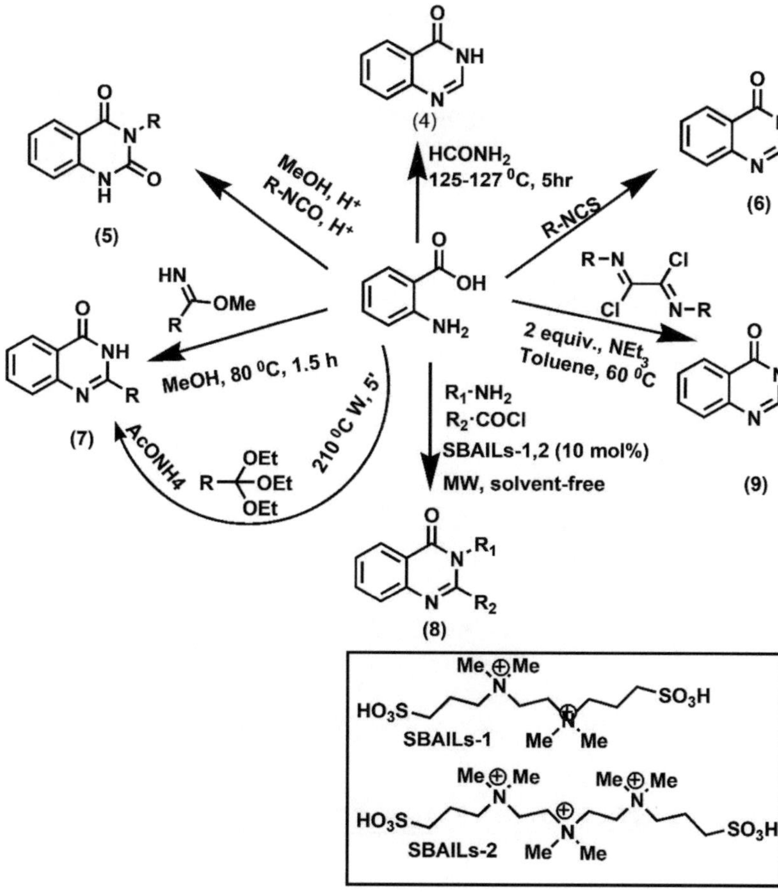

Figure 2.
Synthesis of quinazolin-4(3H) one from 2-aminobenzoic acid.

2.2 Synthesis of quinazolinone compounds from 2-aminobenzamide

In 1962, Bake and Almaula [31] have reported the synthesis of 2-carboethoxy-quinazoline-4(3H)-one **10** through the reaction of anthranilamide and diethyl oxalate (**Figure 3**). Shaterian and Rigi [32] reported a starch sulfate-catalyzed method for synthesis of 2-substituted-1,2,3,4-tetrahydro-4-quinazolinones **11** (**Figure 3**). Zhang and co-workers [33] reported a MnO_2-catalyzed method for the synthesis of 2-substituted quinazolinones **12**. Anthranilamides undergo a-MnO_2-catalyzed oxidative cyclization with alcohols using TBHP as an oxidant (**Figure 3**). Compound **12** could be obtained through the condensation of anthranilamide with an aldehyde in refluxing ethanol in the presence of $CuCl_2$ [34]. Schiff base intermediate was first obtained and, in turn, is transformed into the 2-substituted quinazolinones **12** (**Figure 3**).

In 1887, when Körner reported that the acylation of anthranilamide results in diamide intermediate which upon treatment with sodium carbonate or sodium hydroxide yielded 2-phenylquinazolin-4(3H)-one **12** (**Figure 3**) [35].

Quinazolin-4(3*H*) one compound **12** have been developed by Yang et al. via selective cleavage of the triple bond of ketoalkynes. A reasonable mechanism was suggested for this reaction (**Figure 3**). Michael addition of the amino group of the anthranilamide to the triple bond of the ketoalkyne generated the enaminone intermediate which upon acid catalyzed intramolecular cyclization with subsequent C-C bond cleavage afforded final product **12**.

2.3 Synthesis of quinazolinone compounds from o-substituted aniline

Yan et al. [36] reported a C(sp3)-H oxidative amination, tandem condensation oxidation, catalyzed by iodine method to access quinazolinone compound **13** (**Figure 4**).

Natte and co-workers [37] reported the reaction of 2-iodoanilines, amines and trimethyl orthoformate catalyzed by Pd/C afforded quinazolin-4(3*H*) one compounds **13** (**Figure 4**).

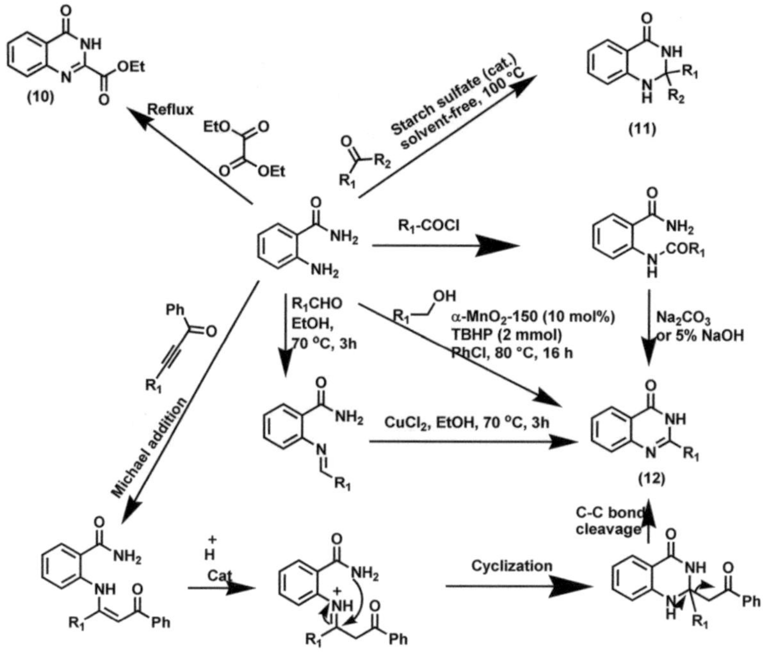

Figure 3.
Synthesis of quinazolin-4(3H) one from 2-aminobenzamide.

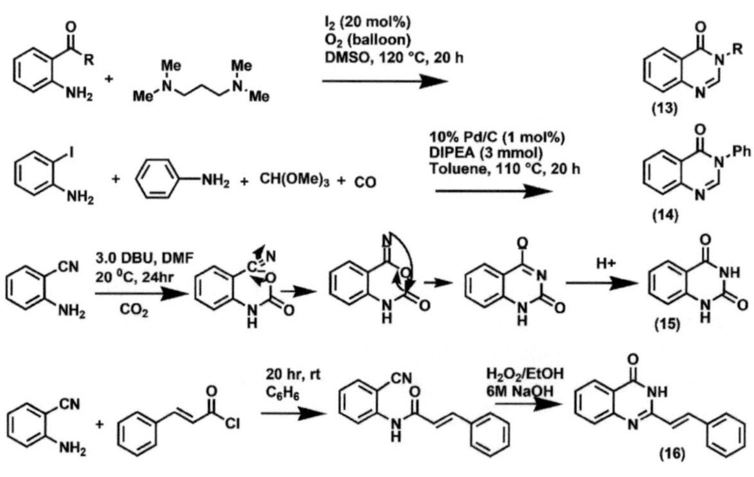

Figure 4.
Synthesis of quinazolinone compounds from o-substituted aniline.

Mizuno et al. [38] synthesis of qiunazolin-2,4-(1*H*,3*H*) ones (2,4-dihydroxy-quinazolines) **15** using 2-aminobenzonitriles starting materials in the presence of carbon dioxide and suitable base. The reaction first generates the carbamate salts followed by nucleophilic cyclisation via attack of the carbamate oxygen onto the cyano group with subsequent rearrangement into an intermediate that is protonated yielding the desired 2,4-dihydroxyquinazoline **15** (**Figure 4**). Also, amidation of anthranilamide with 3-phenylacryloyl chloride with subsequent oxidative ring closure using base catalyst revealed 2-styryl-4(3*H*) quinazolinone **16** (**Figure 4**) [39].

2.4 Synthesis of spiroquinazolinones

Revathy and Lalitha [39] reported a method of *p*-Toluene sulfonic acid-catalyzed synthesis of spiroquinazolinones **17** using anthranilamide and with ketones as starting materials (**Figure 5**). Tajbakhsh et al. [40] reported a H_3PO_3-catalyzed method for synthesis of spiro2,3-dihydroquinazolin-4(1*H*)-ones **18** using isatoic anhydride, hydrazides and cyclic ketones, in the presence of H_3PO_3 catalyst (20 mol %, in ethanol) (**Figure 5**).

2.5 Synthesis of heterocycle-fused quinazolinones

Yang et al. [41] reported the synthesis of tricyclic quinazolinones **19** using formic acid-catalyzed intramolecular cyclization of 3-(2-aminoalkyl)-2-(phenylamino) quinazolin-4(3H)-ones (**Figure 6**).

Reddy et al. [42] reported a CuI/DMSO-catalyzed domino oxidative method for the synthesis of tryptanthrin compound **20** through the interaction of 2-aminoacetophenone and isatoic anhydrides (**Figure 6**).

Foldesi et al. [43] reported the synthesis of the tetracyclic pyrrolotriazepino-quinazolinone derivative **21** the interaction between 1-aryl-4-(methylsulfanyl)-5H-pyrrolo[2,1-d] [1, 2, 5] triazepines and anthranilic acid under reflux in acetic acid (**Figure 6**).

Yuan et al. [44] developed a process to prepare 1H-pyrimido[2,1-b]quinazoline-2,6-dione derivatives **22.** The product was accessed through a tandem aza-Wittig/nucleophilic addition/intramolecular cyclization/isomerization reaction of (E)-methyl 2-((2-azidobenzamido)methyl)-3-phenylacrylate with triphenylphosphine and isocyanates (**Figure 6**).

Wang and Ganesan [45] reported the synthesis of luotonin A, **23** through the reaction of anthranilic acid with 1*H*-pyrrolo[2,3-b]quinolin-2(3*H*)-one (**Figure 6**).

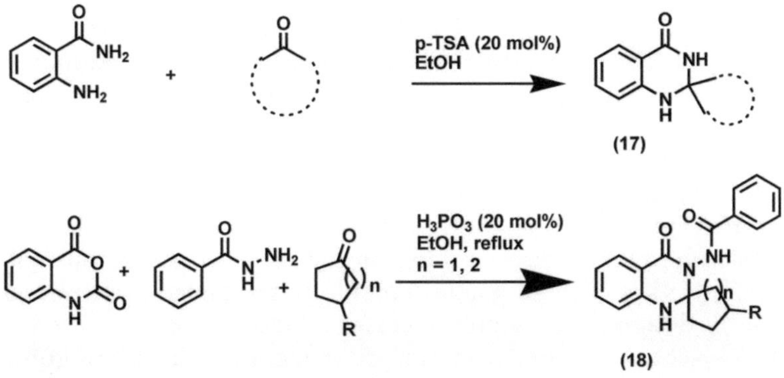

Figure 5.
Synthesis of spiroquinazolinones.

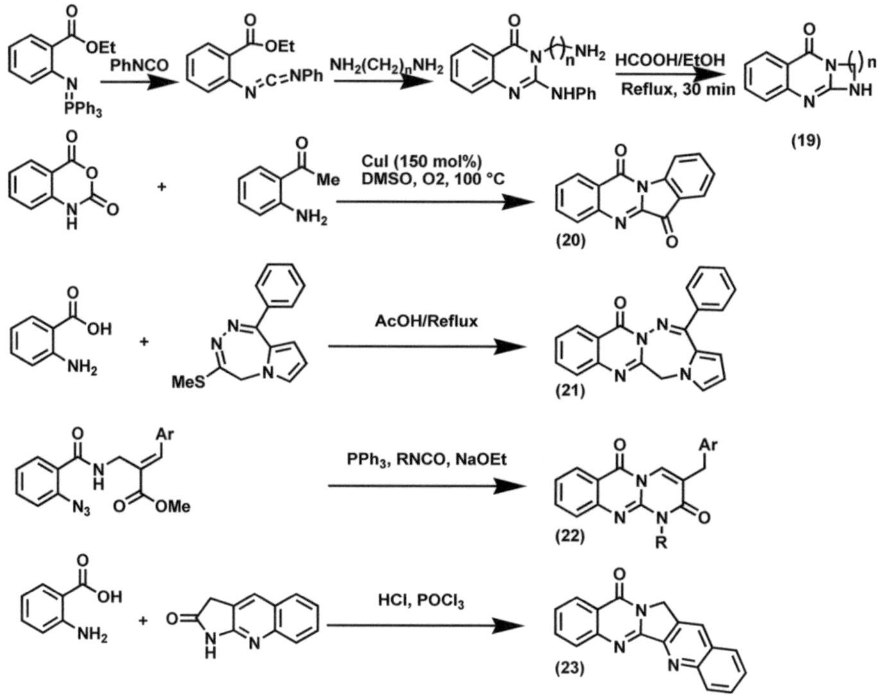

Figure 6.
Synthetic methods of heterocycle-fused quinazolinones.

3. Biological applications of quinazolinones

Natural quinazolinones that widely used in traditional folk medicines are isolated from the plants and microorganisms while the major quinazolinone derivatives are accessed through synthetic process by some chemical reactions. Quinazolinone compounds constitute most privileged class of biologically active heterocyclic compounds. Because of their wide spectrum of biological activities, quinazolinones either from natural source or from synthetic origin, have prompted the medicinal chemist for structural design of these active compounds to develop high selective and potent pharmacological activities.

3.1 Anticancer activity

The natural cytotoxic quinazolinones are depicted in **Figure 7**. The Chinese herbal medicinal plant, Luotonin A **23, Figure 7** is a cytotoxic natural alkaloid possessing pentacyclic fused-quinazolinone moiety. It was first isolated from *Peganum migellastrum* in 1997 and it is in clinical use as anticancer agent and showed low human human topoisomerase-I inhibitor activity [46].

Topoisomerases being major targets for anticancer drug design, the luotonin A was used as a lead compound for development of analogs with increased potency [47]. In comparison with the luotonin A, the majority derivatized analogs explored higher activity for topoisomerase I inhibition and better *in vitro* cytotoxicity than lutonin A [47]. In view of these results, luotonin A is considered as a pharmacophoric core for the design of new topoisomerase I inhibitors [48].

In 2006, (−)-chaetominine **24**, a tetracyclic tripeptide alkaloid (**Figure 7**) was isolated from an endophytic fungus, Chaetomium sp. IFB-E015, and showed smaller IC_{50} than the most frequently prescribed anticancer drug 5-fluorouracil

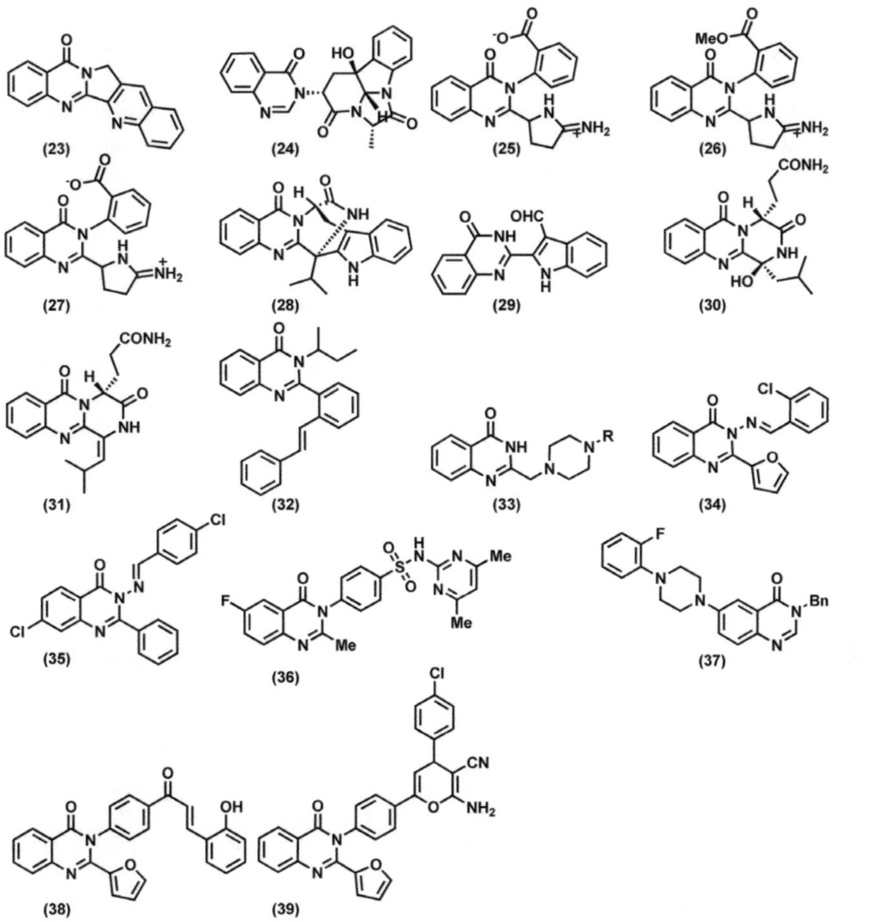

Figure 7.
Chemical structure of anticancer compounds.

against human leukemia K562 (IC50 21 μM) and colon cancer SW1116 (IC50 28 μM) cell lines with IC_{50} values [49].

Quinazolinone alkaloids, auranomides A–C (**25–27**) were isolated from *Penicillium aurantiogriseum* and showed moderate cytotoxicity against human tumor cells [50]. Sartorymensin (**28**), a hexcyclic indoloazepine-fused quinazolinone derivative was isolated from *Neosartorya siamensis* fungus (KUFC 6349) showed a moderate *in vitro* growth inhibitory activity of several cancer cell lines [51]. The 2-substituted quinazolinone alkaloid, bouchardatine (**29**), isolated from *Bouchardatia neurococca* (*Rutaceae*), showed anti-cancer effects [52]. The aurantiomides B and C (**30**, **31**) isolated from *Penicillium aurantiogriseum* fungus derived from sponge and showed moderate activities (IC50 values of 52–54 μg/mL against HL-60 and P-388 for aurantiomide B and IC50 values of 48–62 μg/mL against P-388 and BEL-7402 for aurantiomide C) [53].

Mahdavi et al. [54] developed a new N-substituted 2-arylquinazolinones bearing transstilbene moiety **32** showed good profile (IC50 < 5 μM) against human ductal breast epithelial tumor (T-47D) and human breast adenocarcinoma (MCF-7 and MDA-MB-231) showing two-fold potency more than etoposide standard drug. 2-(piperazin-1-*yl*-methyl)quinazolin-4(3*H*)-one (**33**, R = Acetyl, Propionyl) showed potent anti-cancer activity [55]. While, 3-(2-chloro benzylidinamine)-2-(furan-2-yl)quinazoline-4(3*H*)-one (**34**) showed high activity against ovarian and non-small cell lung cancer [56]. 7-chloro-3-{[(4-chlorophenyl)methylidene] amino}-2-phenylquinazolin-4(3H)-one (**35**), explored significant activity against CNS Cancer cell line [57].

Zayed et al. [58] developed a quinazolinone-bearing sulphonamide moiety compound **36** which upon MTT ass showed IC50 value of 2.51 μM, whereas, the reference drug (methotrexate) exhibited an IC50 value of 2.4 μM, against Michigan Cancer Foundation-7 (MCF-7) breast cancer cells, National Cancer Institute (NCI) lung cancer cells and Human Embryonic Kidney-293 (HEK-293) normal kidney cell. 6-Substituted quinazolinone compound **37** was reported by Malinowski et al., revealed significant activity toward both HT29 (IC50 = 50.90 μM) and HCT116 (IC50 = 46.00 μM) cells lines [59].

Ahmed and Belal reported 2-furylquinazolinone derivatives including compound **38** that explored IC50 value equal to 7 μM/mL on MCF7 cells, and promising inhibitory activity against EGFR-TK [60], and compound **39** depicted a 24-fold higher potency than doxorubicin on HCT116 cancer cells, with IC50 values of 0.2 nmol/mL. Also, compound **39** showed a similar potency to the doxorubicin on MCF7 cell lines and remarkable EGFR inhibitor activity compared with erlotinib standard drug [61].

3.2 Anti-inflammatory activity

The inflammation is a biochemical reactions response that protects the body from infection and injury. It reflects the response of the organism to various stimuli and is related to many disorders such as arthritis, asthma and psoriasis which require prolonged or repeated treatment. The major cause of inflammation the release of chemicals from tissues such as the prostaglandins, histamine, leukotrienes, bradykinin, platelet-activating factor and interleukin-1. Corticosteroids inhibit the synthesis of both PGs and LTs through the release of lipocortin, which inhibits phospholipase A2 and subsequently reduces arachidonic acid release alleviating the inflammation of either rheumatoid arthritis or asthma. While nonsteroidal anti-inflammatory drugs NIASID relieve the inflammation through the inhibition of the cyclooxygenase enzyme and reducing the synthesis of prostanoids [62]. **Figure 8** shows the chemical structure of the anti-inflammatory quinazolinone compounds. *Spiro* [(*2H,3H*) quinazoline-2,10-cyclohexan]-4(*1H*)-one compounds **40** and **41** were reported as potent anti-inflammatory and analgesic activity of superior GIT safety margin in rats model compared with indomethacin (10 mg/kg) and tramadol (20 mg/kg) as reference standards [63].

Abbas SE et al. [64] reported a new quinazolinone-pyrimidine hybrid compound (**42**) which showed more activity and less ulcerogenicity than diclofenac (IC50 = 116.73 μmol/kg; ulcer index = 11.38). The compound explored two-fold more selective inhibition of COX-2 than COX-1.

Hemalath K et al. [65] reported a novel quinazolinone derivative (**43**) that explored 36.3 inhibition of oedema in animal model and showed anti-inflammatory activity as same as phenyl butazone reference drug at a p.o. dose of 25, 50, and 100 mg/kg. Hemalath K et al. [65] also developed a 2,3-dihydroquinazolin-4(1H)-one (**44**) succeeded to produce higher protection against bovine serum albumin (BSA) denaturation that displayed higher protection than diclofenac sodium reference drug. Manivannan et al. [66] designed new quinazolinone derivatives (**45**; R1 = H, Br, R2 = H, Cl; R3 = H, OCH_3) and assayed the derivatives for cyclooxygenase inhibitions by ovine COX and carrageenan-induced rat paw oedema methods. Four compounds showed potent anti-inflammatory activity with oedema inhibition percentage of 49 ± 1.16, 45 ± 0.82, 46 ± 1.36 and 54 ± 1.83 using indomethacin drug as reference.

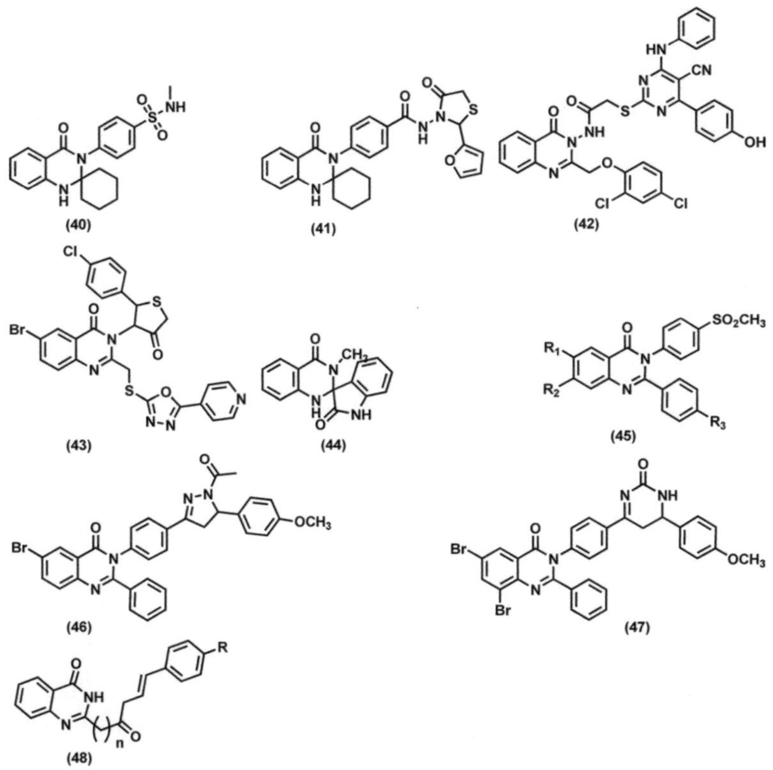

Figure 8.
Chemical structure of anti-inflammatory quinazolinone derivatives.

Mohamed MS et al. [67] have synthesized a series of 2-phenyl-4(3H) quinazolinone derivatives of which to compounds (**46, 47**) showed considerable potent anti-inflammatory activity in rats model with oedema inhibition percentage of 46 ± 1.26, 43 ± 1.82 using indomethacin standard drug in 3 hrs.

Rakesh et al. [68] have synthesized a series of Schiff base derivatives of quinazolinone and got excellent anti-inflammatory activity for synthesized compounds (48; n = 2,3, R = Cl, NO_2).

3.3 Anticonvulsant activity

Epilepsy is defined a chronic neurological syndromes and marked by neuronal firing and neuronal hyperexcitability. Although, the available antiepileptic therapeutics explore satisfactory seizure control in about 70% of epileptic patients, it has become very urgent to search for new antiepileptic compounds with fewer side-effects and less toxicity.

Figure 9 shows chemical structures of anticonvulsant agents. The structure-activity relationship of the anticonvulsant activity of 4(3H)-quinazolinones nucleus revealed the crucial role of a methyl group at position 2 and a substituted aromatic ring at position 3 for the anticonvulsant activity of compounds such as methaqualone (**49**; R1 = CH3, R2 = o-tolyl, R3 = H), etaqualone (49; R1 = CH3, R2 = 2-ethylphenyl, R3 = H), mecloqualone (**49**; R1 = CH3, R2 = o-chlorophenyl, R3 = H), methylmethaqualone (**49**; R1 = CH3, R2 = 2,4 = dimethylphenyl, R3 = H), piriqualone (49; R1 = CH3, R2 = ((E)-2-pyridin-2-ylethenyl), R3 = H) and afloqualone (**49**; R1 = fluoromethyl, R2 = o-tolyl, R3 = NH2) [69]. Against electroshock induced convulsions methaqualone is 1.5 times more potent anticonvulsant than phenytoin sodium and

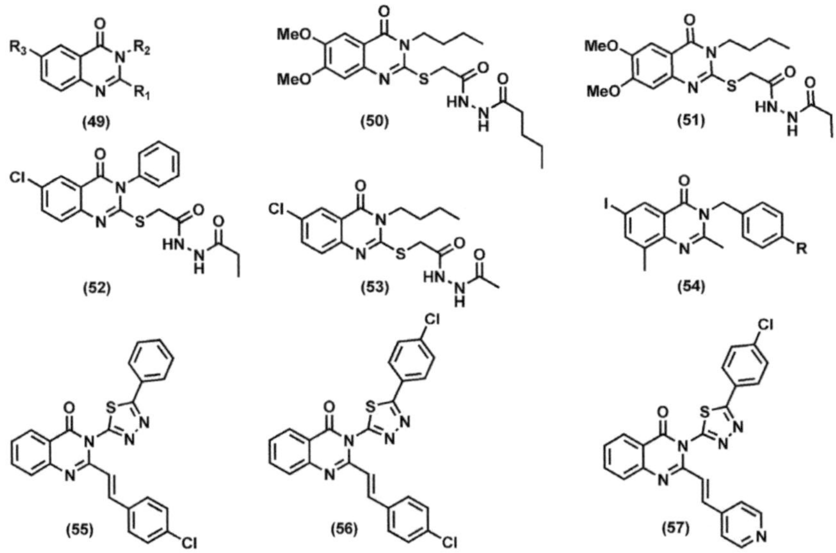

Figure 9.
Chemical structure of anticonvulsant agents.

against pentylenetetrazol-induced seizures it is 10 times more potent than troxidone [70]. Methaqualone produces anticonvulsant effects, through the GABA type A receptors, at low doses while at higher doses, it produces muscle-relaxant and sedative effects [71].

Al-Salem et al. [70] reported 4(3H)-quinazolinone bearing hydrazinecarbothioamide, benzenesulfonohydrazide or phenylacyl acetohydrazide moiety. Compounds **50–53** were most potent with 100% protection against PTZ-induced convulsions compared with the reference drug sodium valproate. Abuelizz HA et al. [72] synthesized a of 6,8-diiodo-2-methyl-3-substituted-quinazolin-4(3H)-ones analogues (**54**; R = Cl, Br, F) that revealed good anticonvulsant activity as evaluated by the maximal electroshock-induced seizure and subcutaneous pentylenetetrazole tests.

A series of novel 3-[5-substituted phenyl-1,3,4-thiadiazole-2-yl]-2-styryl quinazoline-4(3H)-one derivatives has been synthesized by Jatav et al. [73]. Of this series, compounds **55** and **56** revealed anticonvulsant activity results at 0.5 and 4 h in both MES and scPTZ test models, whereas compound **57** explored anticonvulsant activity results at 4 h in MES model and at 0.5 and 4 h in scPTZ model.

3.4 Antimicrobial activity of quniazolinones

2-oxo-azetidinyl-quinazolin-4(3H)-ones (**58**) possess antimicrobial activity against *S. aureus*, *B. subtilis*, *E. coli* and *C. albicans* [74]. 2-Mercapto-3-(4-chlorophenyl)-6-iodo-3H-quinazolin-4-one derivatives (**59–62**) were reported to show a significant antimicrobial activity and could be useful as lead compounds for further design and discovery of more potent antimicrobials [75].

Vani et al. [76] synthesized a series of quinazolin-4(3H)-one-triazole hybrids (**63–65**) quinazolin-4(3H)-one and oxadiazole hybrids (**66–68**). Compounds **64–67** showed significant antibacterial activity against all the bacterial strains, Gram-positive bacteria (*Staphylococcus aureus, Bacillus subtilis*) and Gram-negative bacteria (*Escherichia coli, Pseudomonas aeruginosa*), while compounds (**63, 67, 68**) showed highest activity against fungal species, *Candida albicans* and *Aspergillus niger* compared with ciprofloxacin and fluconazole as reference drugs, respectively.

Kohli et al. reported the antimicrobial activity of the derivatives of 4-Oxo-2-phenylquinazolin-3(4H)-yl-amine (**69**; R = H, CH_3, C_2H_5) against bacterial strains *Staphylococcus aureus* and *Escherichia coli* using ampicillin drug reference at a concentration 100 μg/ml [77].

Sowjanya et al. [78] reported the synthesis of 2-(Substituted styryl)-quinazoline-4(*3H*)-ones (**70**; R = H, CH_3, C_2H_5) as antibacterial agents against bacterial strains *Bacillus subtilis*, *Staphylococcus aureus* (Gram-positive), *Escherichia coli* and *Proteus vulgaris* (Gram-negative) using 100 μg/ml of streptomycin and penicillin as standard drugs.

El-Hashash et al. [79] reported the synthesis of quinazolinones bearing sulfonamide moiety (**71–73**) as antimicrobial against *Gram*-positive bacteria *S. aureus* and *B. cereus* and *Gram*-negative bacteria *S. marcescens* and *P. mirabilis* and as antifungal agents against *A. ochraceus Wilhelm* and *P. chrysogenum* using ampicillin and mycostatin as standards, respectively. **Figure 10** shows the chemical structure of antimicrobial compounds (**58–73**).

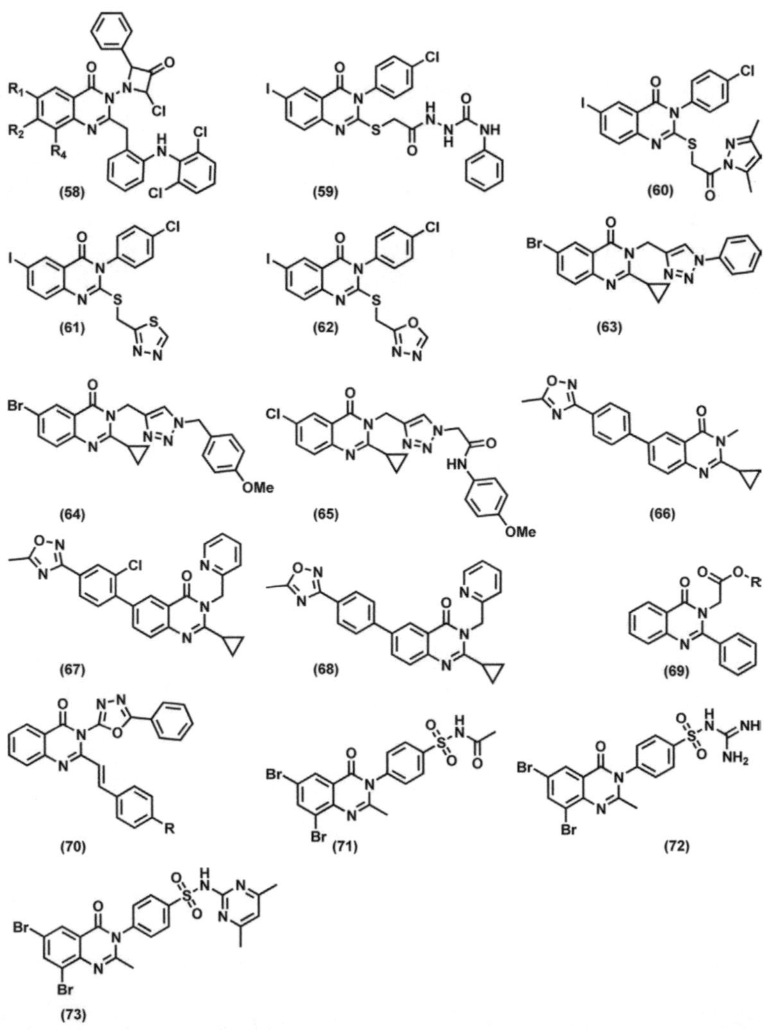

Figure 10.
Chemical structure of antimicrobial compounds.

3.5 Antimalarial activity

Malaria is a parasitic disease caused by *Plasmodium* species parasite. It is widespread in several regions in Africa, Asia and South America. These parasites have developed a drug resistance to almost all the commercially available antimalarial drugs. The good antimalarial potency and the less side effects of quinazolinone compounds promote the researchers for the development of new antimalarial compounds [80].

In 1948, Febrifugine (**74**), a Chinese traditional herb, has been extracted from leaves of *Dichroa febrifuga* that was found in the garden plant Hydrangea. It has 50–100 times as antimalarial as quinine in *in vivo* model. Febrifugine analogues WR140085 (**75**), WR090212 (**76**), WR146115 (**77**) were reported as potent antimalarial agents. The gastrointestinal side effects of (**74**) and the macrophage cells mediated clearance of (**75–77**) requested further therapeutic development and discovery of new antimalarial drugs [2]. Birhan et al. [81] have synthesized 3-aryl-2-(substituted styryl)-4(3*H*)-quinazolinone derivatives (**78, 79**) as potent antimalarial agents (**Figure 11**).

3.6 Antiviral activity

Wang et al. [82] reported quinazolinone derivatives (**80, 81**) with potent antiviral activity against HIV and TMV. Gao et al. [83] have synthesized a series of 2-aryl- or 2-methyl-3-(substituted benzalamino)-4(3*H*)-quinazolinone derivatives and found that the compounds (**82**) and (**83**) exhibit good antiviral activity against TMV (**Figure 12**).

Liu et al. [84] reported a series of 2-pyridinyl-3-substituted-4(3H)-quinazolinones as anti-influenza A virus agents. Of these derivatives, compounds (**84–87**) revealed potent activity (IC50 = 51.6–93.0 μM) better than that of the clinically used drug, ribavirin. Also, it was reported that compound (**87**) could inhibit influenza A virus propagation through inhibition of cellular NF-kB pathway, although it was not as effective as ribvarin (**Figure 12**).

3.7 Cathepsin inhibitor activity

Cathepsins B and H are cysteine proteases that plays a major role in cancer progression as they degrade extracellular matrices facilitating invasion, angiogenesis and metastasis. Therefore the research community has been prompted to the discovery of potent cathepsins inhibitor hemotherapeutics [85].

Singh and Raghav [86] reported the synthesis of a series of 2,3-dihydroquinazolin-4(*1H*)-ones and evaluated it as cathepsins inhibitors, **Figure 13**. Of these

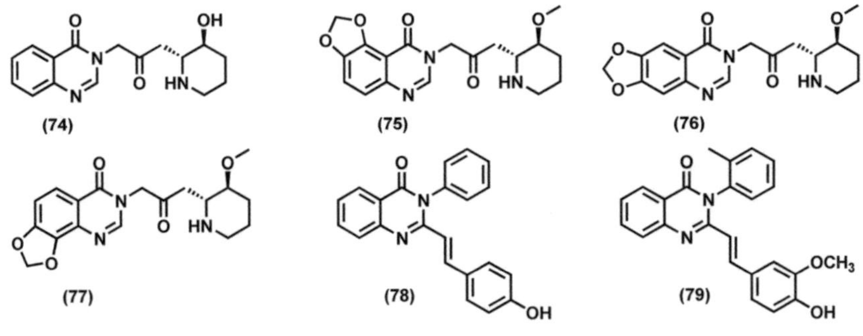

Figure 11.
Antimalarial compounds.

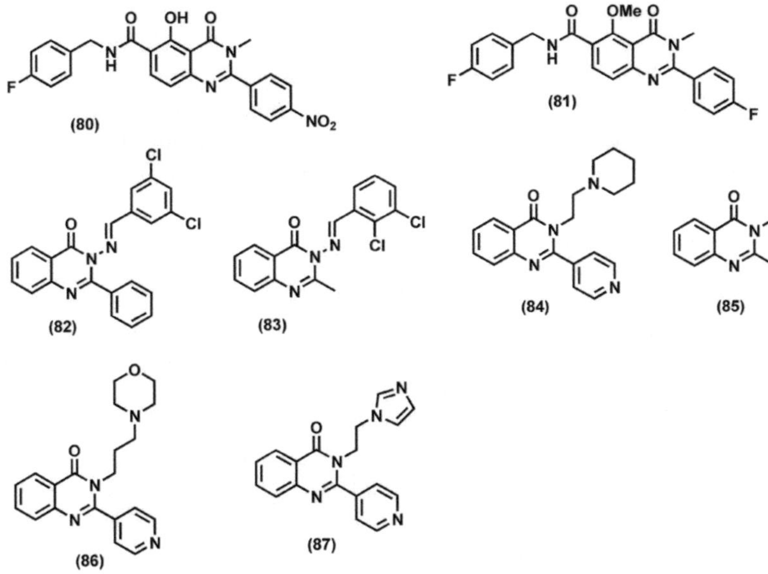

Figure 12.
Antiviral compounds.

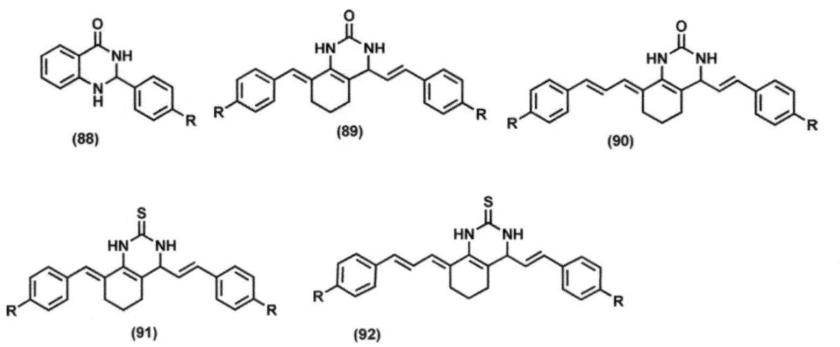

Figure 13.
Chemical structure of cathepsins inhibitors.

compounds, 2-(4-fluorophenyl)-2,3-dihydroquinazolin-4(*1H*)-one (**88**; R = F) and 2-(4-chlorophenyl)-2,3-dihydro quinazolin-4(*1H*)-one (**88**; R = Cl) substituted compounds showed maximum inhibition on cathepsin B. Whereas for cathepsin H, 2,3-dihydro-2-(4-methylphenyl)quinazolin-4(*1H*)-one (**88**; R = Me) and 2-(4-fluorophenyl)-2,3-dihydroquinazolin-4(*1H*)-one (**88**; R = F) have been found to be the most potent inhibitors.

Raghav and Singh [87] reported the synthesis of bischalcones and their quinazoline-2(1*H*)-one (**89, 90**; R = H, NO_2, CH_3O, $(CH_3)_2N$) and quinazoline-2(1*H*)-thione (**91, 92**) derivatives as cathepsin B and cathepsin H inhibitors, (**Figure 13**). Bischalcones and their quinazoline-2(1*H*)-thione derivative (**91, 92**; R = H, NO_2, CH_3O, $(CH_3)_2N$) inhibited both cathepsins in a competitive manner whereas quinazoline-2(1*H*)-one derivative (**89, 90**) inhibited both cathepsins in a non-competitive manner.

3.8 Topoisomerase inhibitor activity

The DNA replication process is controlled essentially by DNA topoisomerase I (Top1) through the relaxation of the nucleic acid's supercoiled structure. Basically,

DNA Top1 attracts the interests of research community as a cancer chemotherapy target [88]. Efforts to overcome side effects of these clinically used anticancer Top1 inhibitors, particularly bladder toxicity, had led to the development of luotonin A alkaloid and discovery of its Top1 inhibitory activity [89].

Ibric et al. reported the development of novel luotonin A isomeric congeners bearing an amino at positions 1, 2, 3, and 4, (**93–97**) **Figure 14** [90]. These compounds revealed significant profile of cytotoxic activity and G2/M cell cycle arrest, proposing either Top1 is not the only target, or some atypical mechanism is accountable for inhibition of Top1 enzyme.

Khadka et al. [91]. synthesized 2-arylquinazolinone derivatives (**98–100**) to investigate these compounds as effective, safe, and selective cytotoxic agents targeting topoisomerases (topos). These compounds showed superior potency as topo I-inhibitors but were inactive against topo IIa.

Kamata et al. have prepared series of pyrimidoacridones (101), Pyrimidocarbazoles (**102**) and pyrimidophenoxadines (**103**) (**Figure 14**), and as topoisomerase II inhibitors [92]. Against P388 and KB cell lines, pyrimidocarbazoles and pyrimidophenoxadines were more potent than pyrimidoacridines. Pyrimidocarbazoles inhibited the *in vivo* tumor growth of mouse sarcoma M5076 with T/C values of 42% at 3.13 mg/kg/d, and increased the level of DNA-topo II cross-linking in P388 cells.

3.9 α-Glucosidase inhibitor activity

Diabetes is a reduced ability to convert glucose into energy inside the body. The role of insulin is the glucose transfer from blood into cells. A large number of antidiabetic agents with different mechanism of action are available in the market.

Saeedi M et al. [93] reported a series of quinazolinone-1,2,3-triazole hybrids 10a-p as potent α-glucosidase inhibitors for use as anti-diabetic agents that exhibited

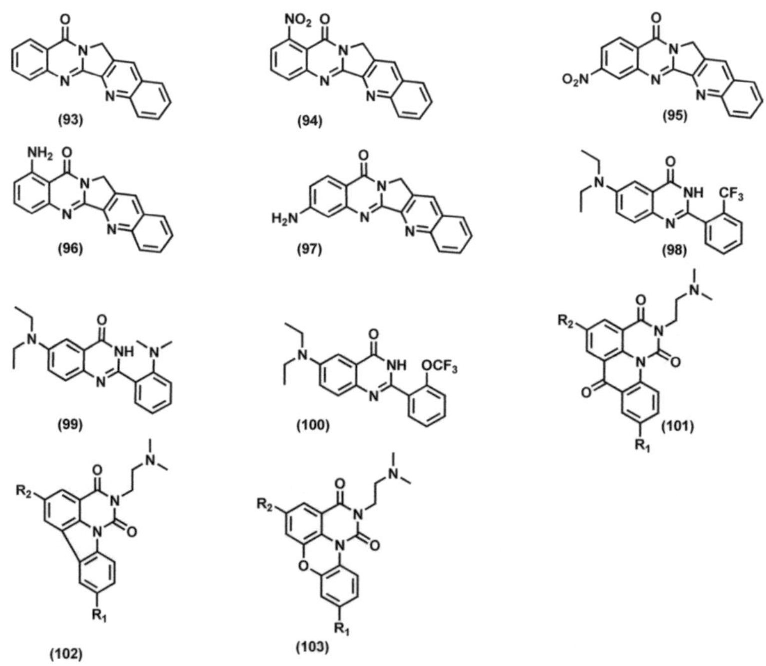

Figure 14.
Chemical structure of topoisomerase enzyme inhibitors.

more potent inhibitory activity against yeast α-glucosidase (IC50 181.0–474.5 μM) than reference drug acarbose (IC50 = 750.0). From these compounds, quinazolinone-1,2,3-triazoles possessing 4-bromobenzyl moiety connected to 1,2,3-triazole ring (**104, 105; Figure 15**) demonstrated the most potent α-glucosidase inhibitor activity. Triazoloquinazoline compounds (**106–110; Figure 15**) were reported as highly potent inhibitor of α-glucosidase enzyme (IC50 = 12.70 ± 1.87, 28.54 ± 1.22, 45.65 ± 4.28, 72.28 ± 4.67 and 83.87 ± 5.12 μM, respectively) compared with the reference standard acarbose (IC50 = 143.54 ± 2.08 μM) [94].

Javaid et al. [95] reported a quinazolinone derivative (**111; Figure 15**) with IC50 value of 0.3 ± 0.01 μM which is about 2800-fold more potent than acarbose reference standard drug. Rahman et al. [96] synthesized a series of hybrid compounds consisting of N-substituted-(4-oxo-2-substituted-phenylquinazolin-3-(4H)-yl) substituted benzene sulfonamide derivatives. From this series, compounds (**112–116; Figure 15**) explored significant antidiabetic activity.

3.10 Thymidine synthase inhibitor activity

Thymidylate synthase enzyme (TS) plays a crucial role in the DNA biosynthesis that it catalyzes the conversion of deoxyuridine monophosphate (dUMP) to deoxythymidine monophosphate (dTMP), one of the nucleotides that constitute the DNA. Inhibition of TS results in imbalance of deoxynucleotides that increase the level of dUMP and finally leads to DNA damage [97]. TS is considered as interesting chemotherapeutic target for treatment of pancreatic, colorectal, ovarian, breast and gastric cancers [98].

In 1998, raltitrexed (**117**) a quinazolin-4-(1*H*)-one compound that has been clinically approved by EMA for treatment of colorectal cancer. Also, pemetrexed

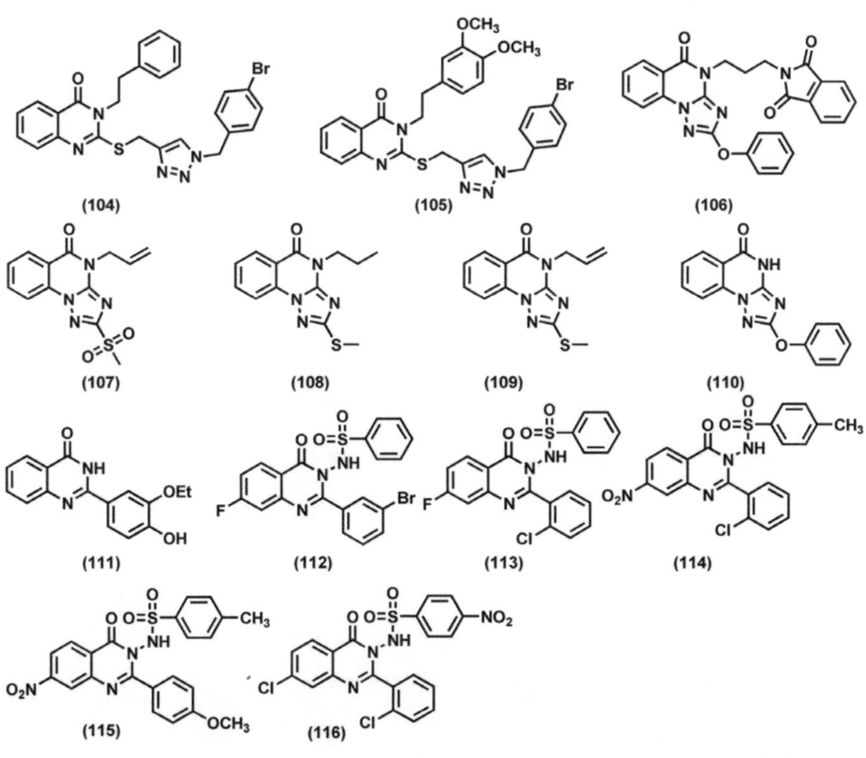

Figure 15.
Chemical structure of compounds with α-glucosidase inhibitor activity.

(**118**) is a quinazolin-4(1*H*)-one compound that is clinically approved by EMA and FDA in 2001. Both raltitrexed and pemetrexed are considered as classical antifolates as they are folate analogs containing a pterin ring and a charged glutamate tail, therefore they need active internalization into the cells through folate carrier system [99].

In recent studies, El-Messery SM et al. [100] synthesized a series of 2,3,6-substituted quinazolin-4(3H)-ones. Of these compounds, compound (**119**; **Figure 16**) is the most potent inhibitor of bovine liver DHFR (with IC50 = of 0.02 μM). It was reported that 2-mercaptoquinazolin-4(3*H*)-one derivative is a potent bovine liver DHFR inhibitor (**120**; **Figure 16**) (IC50 = 0.01 μM) [101]. Javaid and co-workers [102] reported a series of 25 2-arylquinazolin-4(3*H*)-ones as potent thymidine phosphorylase inhibitors among these derivatives compounds (**121**, **122**; **Figure 16**) were identified as the lead compounds (IC50 = 42.9 ± 1.0 and 59.5 ± 1.9 μM, respectively).

3.11 Monoamine oxidase inhibitor activity (MAO)

In human, monoamine oxidases (MAOs) are mitochondrial bound enzymes that are responsible for oxidative deamination metabolism of neurotransmitters such as dopamine, serotonin, norepinephrine and epinephrine. In the brain, MAO-A enzyme isoform metabolizes serotonin, therefore specific MAO-A inhibitors are used the treatment of anxiety and depression disorder [103]. On the other hand, MAO-B enzyme metabolizes dopamine in the brain thus MAO-B specific inhibitors are prescribed for the treatment of Parkinson's disease [104].

Quinazolinone moiety, one of numerous MAO inhibitor scaffolds, it has been explored as lead for the further development of potent MAO inhibitors. Compounds (**123–127**; **Figure 17**) are representative MAO-inhibitor examples of 4(3H)-quinazolinones [105].

Qhobosheane et al. reported seven quinazolinone compounds (IC50 < 1 μM) ascertained as potent and specific MAO-B inhibitors, among them the most potent inhibitor, 2-[(3-iodobenzyl)thio]quinazolin-4(3H)-one (**128**; **Figure 17**), with IC50 value of 0.142 μM. Although these derivatives have been proved as reversible and competitive MAO-B inhibitor (Ki = 0.068 μM), none of the them were MAO-A inhibitors [106].

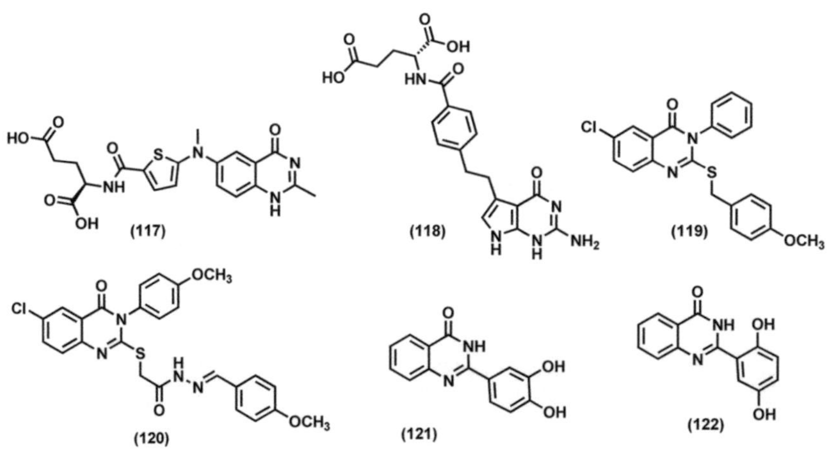

Figure 16.
Thymidine synthase and phosphorylase inhibitors.

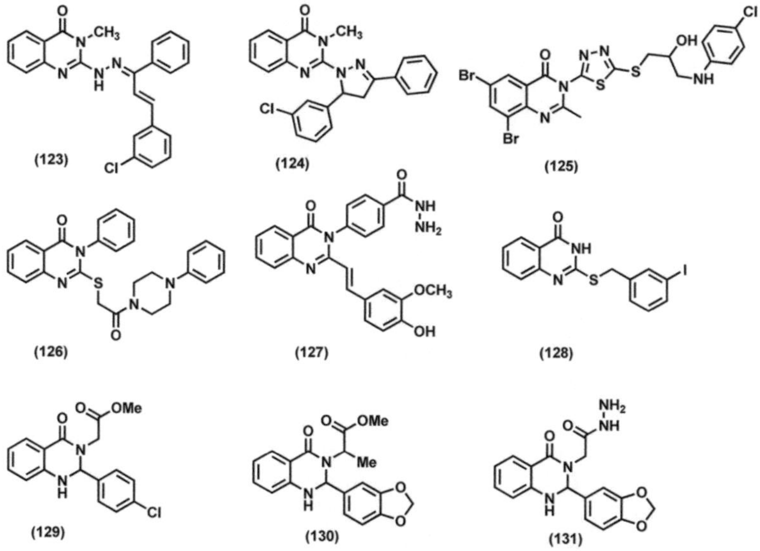

Figure 17.
Chemical structure of monoamine oxidase (MAO) inhibitors.

Khattab et al. [107] reported a series of quinazolinone bearing amino acid ester or amino acid hydrazides (**129–131**; **Figure 17**) that revealed competitive higher inhibitory activity toward MAO-A than MAO-B. The anti-MAO-A activity were comparable with that of the standard reference clorgyline (IC50 = 2.9×10^{-9} M). Compounds (**130, 131**) were the most selective MAO-A inhibitors with selectivity index of **131** (SI = 39,524) superior to that of the reference drug clorgyline (SI = 33,793).

4. Marketed quinazolinone drugs

Proquazone is non-steroidal anti-inflammatory drug (Biarison®) (**132**; **Figure 18**) manufactured by Novartis pharmaceutical company. Also, it is used in the treatment of degenerative joint disease.

Nolatrexed [108] compound (**133**; **Figure 18**) is a thymidylate synthase inhibitor drug manufactured by Agouron pharmaceutical company under trade name Thymitaq®. In 1998, Zarix licensed Thymitaq® from Agouron. It is used in treatment of liver cancer.

Quinethazone or Hyromox® [109] (**134**; **Figure 18**) has been marketed as antihypertensive drug by Lederle pharmaceutical company and was recently withdrawn from the market.

Fenquizone's brand name is Idrolone® (**135**; **Figure 18**) [110], it is marketed by Maggioni pharmaceutical company. It is a low-ceiling diuretics used in the treatment of oedema and hypertension.

The brand name of albaconazole [111] is Albaconazole® (**136**; **Figure 18**). It was marketed by GlaxoSmithKlyne pharmaceutical company as an oral and topical antifungal agent.

Febrifugine, Dichroin B® or ChangShan® (**137**; **Figure 18**) [112] was isolated from Chinese herb Dichroa febrifuga as potent antimalarial drug and was marketed by Hawaii Pharm pharmaceutical company.

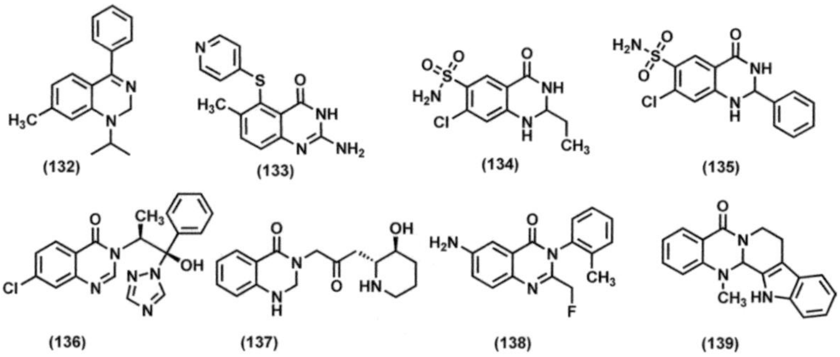

Figure 18.
Chemical structure of marketed quinazolinone drugs.

The brand name of afloqualone is Arofuto® (**138**; **Figure 18**) [113]. It is marketed by Mitsubishi Tanabe Pharma pharmaceutical company as sedative and muscle relaxant drug.

Evodiamine (**139**; **Figure 18**) [114] has been isolated from the Evodia plants and was found to reduce fat uptake in animal model. It is used for bodybuilding as over the counter supplements.

5. Conclusions

This chapter depicts different methods of synthesis of quinazolinone derivatives starting from affordable and easily accessible substrates including 2-aminobenzoic acid, 2-aminobenzamide, o-substituted aniline in addition to the synthetic methods of spiroquinazolinones and heterocycle-fused quinazolinones. Also, the chapter discusses different biological applications of both natural and synthetic quinazolinones. The last section in this chapter lists common quinazolinone drugs that have been approved in the market.

Acknowledgements

This book chapter was supported by King Saud University, Vice Deanship of Research Chairs, Kayyali Chair for Pharmaceutical Industry, through grant number AW-2019.

Conflict of interest

The authors declare no conflict of interest.

Author details

Awwad A. Radwan[1,2*] and Fars K. Alanazi[1,3]

1 Kayyali Chair, College of Pharmacy, King Saud University, Riyadh, Saudi Arabia

2 Department of Pharmaceutical Organic Chemistry, Faculty of Pharmacy, Assiut University, Egypt

3 Department of Pharmaceutics, College of Pharmacy, King Saud University, Riydh, Saudi Arabia

*Address all correspondence to: aradwan@ksu.edu.sa; dhna_2001@hotmail.com

References

[1] Polshettiwar V, Varma RS. Greener and sustainable approaches to the synthesis of pharmaceutically active heterocycles. Current Opinion in Drug Discovery & Development. 2007;**10**:723-737. PMID: 17987524

[2] Jiang S, Zeng Q, Gettayacamin M, Tungtaeng A, Wannaying S, Lim A, et al. Antimalarial activities and therapeutic properties of febrifugine analogs. Antimicrobial Agents and Chemotherapy. 2005;**49**(3):1169-1176. DOI: 10.1128/AAC.49.3.1169-1176.2005

[3] Ghorab MM, Ismail ZH, Abdalla M, Radwan AA. Synthesis, antimicrobial evaluation and molecular modelling of novel sulfonamides carrying a biologically active quinazoline nucleus. Archives of Pharmacal Research. 2013;**36**(6):660-670. DOI: 10.1007/s12272-013-0094-6

[4] Ghoraba MM, Ismail ZH, Radwan AA, Abdalla M. Synthesis and pharmacophore modeling of novel quinazolines bearing a biologically active sulfonamide moiety. Acta Pharmaceutica. 2013;**63**:1-18. DOI: 10.2478/acph-2013-0006

[5] Devi K, Kachroo M. Synthesis and antitubercular activity of some new 2, 3-disubstituted quinazolinones. Der Pharma Chemica. 2014;**6**:353-359. Available from: http://derpharmachemica.com/archive.html

[6] El-Azab AS, El-Tahir KE. Design and synthesis of novel 7-aminoquinazoline derivatives: Antitumor and anticonvulsant activities. Bioorganic & Medicinal Chemistry Letters. 2012;**22**:1879-1885. PMID: 22326394

[7] Radwan AA, Alanazi FK, Al-Dhfyan A. Synthesis, and docking studies of some fusedquinazolines and quinazolines carrying biological active isatin moiety as cell cycle inhibitors of breast cancer cell lines. Drug Research (Arzneimittelforschung). 2013;**63**:129-136. DOI: 10.1055/s-0032-1333306

[8] Khan I, Zaib S, Batool S, Abbas N, Ashraf Z, Saeed A. Quinazolines and quinazolinones as ubiquitous structural fragments in medicinal chemistry: An update on the development of synthetic methods and pharmacological diversification. Bioorganic & Medicinal Chemistry. 2016;**24**:2361-2381. DOI: 10.1016/j.bmc.2016.03.031. Epub 2016 March 18

[9] Malamas MS, Millen J. Quinazolineacetic acids and related analogs as aldose reductase inhibitors. Journal of Medicinal Chemistry. 1991;**34**:1492-1502. DOI: 10.1021/jm00108a038

[10] Priya MGR, Girija K, Ravichandran N. Invitro study of anti-inflammatory and antioxidant activity of 4-(3h)-quinazolinone derivatives. Rasayan Journal of Chemistry. 2011;**4**:418-424. Available from: rasayanjournal.co.in/vol-4/issue-2/28.pdf

[11] Yousefi B, Azimi A, Majidinia M, Shafiei-Irannejad V, Badalzadeh R, Baradaran B, et al. Balaglitazone reverses P-glycoproteinmediated multidrug resistance via upregulation of PTEN in a PPARγdependent manner in leukemia cells. Tumor Biology. 2017;**39**:1-11. DOI: 10.1177/1010428317716501

[12] Traxler P, Green J, Mett H, Séquin U, Furet P. Use of a pharmacophore model for the design of EGFR tyrosine kinase inhibitors: Isoflavones and 3-phenyl-4(1H)-quinolones. Journal of Medicinal Chemistry. 1999;**42**:1018-1026. DOI: 10.1021/jm980551o

[13] Sahu M, Nerkar AG. In silico screening, synthesis and *in vitro* evaluation of some quinazolinone

derivatives as dihydrofolate reductase inhibitors for anticancer activity. International Journal of Pharmacy and Pharmaceutical Sciences. 2014;**6**:193-199. Available from: https://pdfs.semanticscholar.org/39de/8f154b7087ca23349b80a8aed6804c988e6a.pdf

[14] Örfi L, Kökösi J, György S, Kövesdi I, Mak M, Teplan I, et al. Heterocondensed quinazolinones: Synthesis and protein-tyrosine kinase inhibitory activity of 3,4-dihydro-1H,6H-[1,4]oxazino[3,4-b]quinazolin-6-one derivatives. Bioorganic & Medicinal Chemistry. 1996;**4**:547-551. DOI: 10.1016/0968-0896(96)00040-5

[15] Hour MJ, Huang LJ, Kuo SC, Xia Y, Bastow K, Nakanishi Y, et al. 6-Alkylamino- and 2,3-dihydro-3′-methoxy-2-phenyl-4-quinazolinones and related compounds: Their synthesis, cytotoxicity, and inhibition of tubulin polymerization. Journal of Medicinal Chemistry. 2000;**43**:4479-4487. DOI: 10.1021/jm000151c

[16] Bouley R, Ding D, Peng Z, Bastian M, Lastochkin E, Song W, et al. Structure-activity relationship for the 4(3H)-quinazolinone antibacterials. Journal of Medicinal Chemistry. 2016;**59**(10):5011-5021. DOI: 10.1021/acs.jmedchem.6b00372

[17] Bojarski AJ, Kowalski P, Kowalska T, Duszynska B, Charakchieva-Minol S, Tatarczynska E, et al. Synthesis and pharmacological evaluation of new Arylpiperazines. 3-{4-[4-(3-chlorophenyl)-1-piperazinyl] butyl}-quinazolidin-4-one—A dual serotonin 5-HT1A/5-HT2A receptor ligand with an anxiolytic-like activity. Bioorganic & Medicinal Chemistry. 2002;**10**:3817-3827. 10.1016/S0968-0896(02)00349-8

[18] Grosso JA, Nichols DE, Kohli JD, Glock D. Synthesis of 2-(alkylamino)-5,6- and-6,7-dihydroxy-3,4-dihydroquinazolines and evaluation as potential dopamine agonists. Journal of Medicinal Chemistry. 1982;**25**:703-708. DOI: 10.1021/jm00348a018

[19] Mohamed YA, El-galil A, Amrb C, Mohamed SF, Abdalla MM, Al-omar M, et al. Cytotoxicity and anti-HIV evaluations of some new synthesized quinazoline and thioxopyrimidine derivatives using 4-(thiophen-2-yl)3,4,5,6tetrahydrobenzo[h]quinazoline-2(1H)-thione as synthon. Journal of Chemical Sciences. 2012;**124**:693-702. DOI: 10.1007/s12039-012-0242-4

[20] Wattanapiromsakul C, Forster PI, Waterman PG. Alkaloids and limonoids from *Bouchardatia neurococca*: Systematic significance. Phytochemistry. 2003;**64**:609-615. DOI: 10.1016/S0031-9422(03)00205-X

[21] Asif M. Chemical characteristics, synthetic methods, and biological potential of quinazoline and quinazolinone derivatives. International Journal of Medicinal Chemistry. 2014;**2014**:1-27. DOI: 10.1155/2014/39563

[22] Kshirsagar UA. Recent developments in the chemistry of quinazolinone alkaloids. Organic & Biomolecular Chemistry. 2015;**13**(36):9336-9352. DOI: 10.1039/C5OB01379H

[23] Tee OS, Patil GV. Mechanism of bromination of 4(3H)-quinazolinone, its 3-methyl and its 1, 3-dimethyl derivatives in aqueous acidic solutions. The Journal of Organic Chemistry. 1976;**41**:838-845. DOI: 10.1021/jo00867a020

[24] Lowe JA, Archer RL, Chapin DS, Cheng B, Helweg D, Johnson JL, et al. Structure-activity relationship of quinazolinedione inhibitors of calcium-independent phosphodiesterase. Medicinal Chemistry. 1991;**34**:624-628. DOI: 10.1021/jm00106a024

[25] Ried W, Stephan W. Über eine neue synthese von substituierten 4-hydroxy-6.7-benzo-chinazolinen. Chemische Berichte. 1963;**96**:1218-1220. DOI: 10.1002/cber.19630960507

[26] Ried W, Stephan W. Über die Reaktion von freier Anthranilsäure mit freien Imidsäureestern. Chemische Berichte. 1962;**95**:3042-3047. DOI: 10.1002/cber.19620951230

[27] Ried W, Sinharray A. Über Nitroacetimidsäure-ester. Chemische Berichte. 1963;**96**:3306-3311. DOI: 10.1002/cber.19630961230

[28] Rad-Moghadam K, Mohseni M. An expeditious and solvent-free route to the synthesis of 2-substituted quinazolin-4(3H)-ones under microwave conditions. Journal of Chemical Research. 2003;**2003**:487-488. DOI: 10.3184/030823403103174632

[29] Li X, Lin Q, Wang L. One-pot solvent-free synthesis of 2, 3-disubstituted 4(3H)-quinazolinones catalyzed by long-chain double SO3H-functionalized Brønsted acidic ionic liquids under microwave irradiation. Journal of the Iranian Chemical Society. 2015;**12**:897-901. DOI: 10.1007/s13738-014-0553-0

[30] Wuckelt J, Döring M, Beckert R, Langer P. Efficient synthesis of quinazolin-4-ones and axially chiral 2,2′-bisquinazolin-4-ones by reaction of anthranilic acid derived nucleophiles with oxalic acid-bis(imidoyl)chlorides. Synlett. 1999;**7**:1100-1102. DOI: 10.1055/s-1999-2760

[31] Baker BR, Almaula PI. Nonclassical antimetabolites. X.1,2 a facile synthesis of 4-quinazolone-2-carboxylic acid and the structure of Bogert's ammonium salt. The Journal of Organic Chemistry. 1962;**27**:4672-4674. DOI: 10.1021/jo01059a524

[32] Shaterian HR, Rigi F. An efficient synthesis of quinazoline and xanthene derivatives using starch sulfate as a biodegradable solid acid catalys. Research on Chemical Intermediates. 2015;**41**:721-738. DOI: 10.1007/s11164-013-1223-z

[33] Zhang Z, Wang M, Zhang C, Zhang Z, Lua J, Wang F. The cascade synthesis of quinazolinones and quinazolines using an α-MnO_2 catalyst and tert-butyl hydroperoxide (TBHP) as an oxidant. Chemical Communications. 2015;**51**:9205-9207. DOI: 10.1039/C5CC02785C

[34] Guo S, Li Y, Li T, Zhang W, Fan X. ChemInform abstract: Rapid assembly of quinazolinone scaffold via copper-catalyzed tandem reaction of 2-bromobenzamides with aldehydes and aqueous ammonia: Application to the synthesis of the alkaloid tryptanthrin. RSC Advances. 2014;**4**:59289-59296. DOI: 10.1039/C4RA10799C

[35] Connolly DJ, Cusack D, OSullivan TP, Guiry PJ. Synthesis of quinazolinones and quinazolines. Tetrahedron 2005;**61**:10153-10202. DOI: 10.1016/j.tet.2005.07.010

[36] Yan Y, Xu Y, Niu B, Xie H, Liu Y. I2-catalyzed aerobic oxidative C(sp3)–H amination/C–N cleavage of tertiary amine: Synthesis of quinazolines and quinazolinones. The Journal of Organic Chemistry. 2015;**80**:5581-5587. DOI: 10.1021/acs.joc.5b00474

[37] Natte K, Neumann H, Wu X-F. Pd/C as an efficient heterogeneous catalyst for carbonylative four-component synthesis of 4(3H)-quinazolinones. Catalysis Science & Technology. 2015;**5**:4474-4480. DOI: 10.1039/C5CY00907C

[38] Mizuno T, Okamoto N, Ito T, Miyata T. Synthesis of 2,4-dihydroxyquinazolines using carbon dioxide in the presence of DBU under mild conditions. Tetrahedron Letters. 2000;**41**:1051-1053. DOI: 10.1016/S0040-4039(99)02231-5

[39] Revathy K, Lalitha A. p-TSA-catalyzed synthesis of spiroquinazolinones. Journal of the Iranian Chemical Society. 2015;**12**:2045-2049. DOI: 10.1007/s13738-015-0680-2

[40] Tajbakhsh M, Ramezanpour S, Balalaie S, Bijanzadeh HR. Novel one-pot three-component reaction for the synthesis of functionalized spiroquinazolinones. Journal of Heterocyclic Chemistry. 2015;**52**:1559-1564. DOI: 10.1002/jhet.2138

[41] Yang X, Wu M, Sun S, Huang C, Guo H, Wang J, et al. Synthesis of tricyclic quinazolinones via intramolecular cyclization of 3-(2-aminoalkyl)-2-(phenylamino) quinazolin-4(3H)-ones. Molecular Diversity. 2016;**20**:551-556. DOI: 10.1007/s11030-015-9636-9

[42] Reddy BVS, Reddy DM, Reddy GN, Reddy MR, Reddy VK. Domino oxidative cyclization of 2-amino-acetophenones for the one-pot synthesis of tryptanthrin derivatives. European Journal of Organic Chemistry. 2015;**2015**:8018-8022. DOI: 10.1002/ejoc.201501079

[43] Foldesi T, Dancsó A, Simig G, Volk B, Milen M. Synthesis of a new tetracyclic ring system: Pyrrolotriazepinoquinazolinone derivatives. Tetrahedron. 2015;**71**:6759-6763. DOI: 10.1016/j.tet.2015.07.033

[44] Yuan D, Kong H-H, Ding M-W. New efficient synthesis of 1H-pyrimido[2,1-b]quinazoline-2,6-diones via a tandem aza-Wittig/nucleophilic addition/intramolecular cyclization/isomerization reaction starting from the Baylis–Hillman adducts. Tetrahedron. 2015;**71**:419-423. DOI: 10.1016/j.tet.2014.12.006

[45] Wang H, Ganesan A. Total synthesis of the cytotoxic alkaloid luotonin a. Tetrahedron Letters. 1998;**39**:9097-9098. DOI: 10.1016/S0040-4039(98)02004-8

[46] Venkatesh R, Ramaiah MJ, Gaikwad HK, Janardhan S, Bantu R, Nagarapu L, et al. Luotonin-a based quinazolinones cause apoptosis and senescence via HDAC inhibition and activation of tumor suppressor proteins in HeLa cells. European Journal of Medicinal Chemistry. 2015;**94**:87-101. DOI: 10.1016/j.ejmech.2015.02.057

[47] González-Ruiz V, Pascua I, Fernández-Marcelo T, Ribelles P, Bianchini G, Sridharan V, et al. B-ring-aryl substituted luotonin a analogues with a new binding mode to the topoisomerase 1-DNA complex show enhanced cytotoxic activity. PLoS One. 2014;**9**:e95998. DOI: 10.1371/journal.pone.0095998

[48] Cagir A, Jones SH, Eisenhauer BM, Gao R, Hecht SM. Synthesis and biochemical properties of E-ring modified luotonin A derivatives. Bioorganic & Medicinal Chemistry Letters. 2004;**2004**(14):2051-2054. DOI: 10.1016/j.bmcl.2004.02.069

[49] Jiao RH, Xu S, Liu JY, Ge HM, Ding H, Xu C, et al. Chaetominine, a cytotoxic alkaloid produced by endophyticchaetomiumsp. IFB-E015. Organic Letters. 2006;**8**:5709-5712. DOI: 10.1021/ol062257t

[50] Song F, Ren B, Yu K, Chen C, Guo H, Yang N, et al. Quinazolin-4-one coupled with pyrrolidin-2-iminium alkaloids from marine-derived fungus penicillium aurantiogriseum. Marine Drugs. 2012;**10**:1297-1306. DOI: 10.3390/md10061297

[51] Buttachon S, Chandrapatya A, Manoch L, Silva A, Gales L, Bruyère C, et al. Sartorymensin, a new indole alkaloid, and new analogues of tryptoquivaline and fiscalins produced by Neosartorya siamensis (KUFC 6349). Tetrahedron. 2012;**68**:3253-3262. DOI: 10.1016/j.tet.2012.02.024

[52] Rao Y, Liu H, Gao L, Yu H, Tan JH, Ou TM, et al. Discovery of natural

alkaloid bouchardatine as a novel inhibitor of adipogenesis/lipogenesis in 3T3-L1 adipocytes. Bioorganic & Medicinal Chemistry. 2015;**23**:4719-4727. DOI: 10.1016/j.bmc.2015.05.057

[53] Xin ZH, Fang Y, Du L, Zhu T, Duan L, Chen J, et al. Aurantiomides A-C, quinazoline alkaloids from the sponge-derived fungus Penicillium aurantiogriseum SP0-19. Journal of Natural Products. 2007;**70**:853-855. DOI: 10.1021/np060516h

[54] Mahdavi M, Pedrood K, Safavi M, Saeedi M, Pordeli M, Ardestani SK, et al. Synthesis and anticancer activity of N-substituted 2-arylquinazolinones bearing trans-stilbene scaffold. European Journal of Medicinal Chemistry. 2015;**95**:492-499. DOI: 10.1016/j.ejmech.2015.03.057

[55] Murahari MS, Prakash RSJ, Kar SS, Kumar TG, Raj VP, Raju DS. Synthesis, characterization and in vitro anti-cancer activity of quinazolinone derivatives. Journal of Pharmacy Research. 2012;**5**:2743-2746. Available from: jprsolutions.info/newfiles/journal-file-56b9f4b60d7c57.58954497.pdf

[56] Noolvi MN, Patel HM, Bhardwaj V, Chauhan A. Synthesis and in vitro antitumor activity of substituted quinazoline and quinoxaline derivatives: Search for anticancer agent. European Journal of Medicinal Chemistry. 2011;**46**:2327-2346. Available from: http://dx.doi.org/10.1016/j.ejmech.2011.03.015. PMid: 21458891

[57] Noolvi MN, Patel HM. Synthesis, method optimization, anticancer activity of 2, 3, 7-trisubstituted quinazoline derivatives and targeting EGFR-tyrosine kinase by rational approach: 1st cancer update. Arabian Journal of Chemistry. 2013;**6**:35-48. DOI: 10.1016/j.arabjc.2010.12.031

[58] Zayed MF, Ahmed HEA, Ihmaid S, Omar A-SM, Abdelrahim AS. Synthesis and screening of some new fluorinated quinazolinone–Sulphonamide hybrids as anticancer agents. Journal of Taibah University Medical Sciences. 2015;**10**:333-339. DOI: 10.1016/j.jtumed.2015.02.007

[59] Malinowski Z, Fornal E, Nowak M, Kontek R, Gajek G, Borek B. Synthesis and biological evaluation of some amino- and sulfanyl-3H-quinazolin-4-one derivatives as potential anticancer agents. Monatshefte fuer Chemie. 2015;**146**:1723-1731. DOI: 10.1007/s007

[60] Ahmed MF, Belal A. Design, synthesis, and molecular docking studies of 2-(furan-2-yl)-quinazolin-4-one derivatives as potential antiproliferative agents. Archiv der Pharmazie-Chemistry in Life Sciences. 2015;**348**:487-497. DOI: 10.1002/ardp.201400468

[61] Ahmed MF, Belal A. Synthesis, characterization, and biological evaluation of new quinazolin-4-one derivatives hybridized with pyridine or pyran moiety. Research on Chemical Intermediates. 2016;**42**:659-671. DOI: 10.1007/s11164-015-2048-8

[62] Vane J, Botting R. Inflammation and the mechanism of action of anti-inflammatory drugs. The FASEB Journal. 1987;**1**(2):89-96. PMID: 3111928

[63] Amin KM, Kamel MM, Anwar MM, Khedr M, Syamb YM. Synthesis, biological evaluation and molecular docking of novel series of spiro [(2H,3H) quinazoline-2,10-cyclohexan]-4(1H)-one derivatives as anti-inflammatory and analgesic agents. European Journal of Medicinal Chemistry. 2010;**45**:2117-2131. Available from: http://dx.doi.org/10.1016/j.ejmech.2009.12.078. PMid: 20137837

[64] Abbas SE, Awadallah FM, Ibrahin NA, Said EG, Kamel GM. New quinazolinoneepyrimidine hybrids: Synthesis, anti-inflammatory, and

ulcerogenicity studies. European Journal of Medicinal Chemistry. 2012;**53**:141-149. DOI: 10.1016/j.ejmech.2012.03.050

[65] Hemalath K, Madhumith G, Vasavi CS, Munusami PJ. 2, 3-Dihydroquinazolin-4 (1H)-ones: Visible light mediated synthesis, solvatochromism and biological activity. Journal of Photochemistry and Photobiology B: Biology 2015;**143**:139-147. DOIr: 10.1016/j.jphotobiol.2014.12.028

[66] Manivannan E, Chaturvedi SC. Analogue-based design, synthesis and molecular docking analysis of 2,3-diaryl quinazolinones as non-ulcerogenic anti-inflammatory agents. Bioorganic & Medicinal Chemistry. 2011;**19**:4520-4528. DOI: 10.1016/j.bmc.2011.06.019

[67] Mohamed MS, Kamel MM, Kassem EMM, Abotaleb N, Nofal SM, AhmedMF.Novel3-(p-Substitutedphenyl)-6-bromo-4 (3H)-quinazolinone derivatives of promising anti-inflammatory and analgesic properties. Acta Poloniae Pharmaceutica. 2009;**66**(5):487-500. PMID: 19894645

[68] Rakesh KP, Ramesh S, Kumar HMM, Chandan S, Gowda DC. Quinazolinones linked amino acids derivatives as a new class of promising antimicrobial, antioxidant and anti-inflammatory agents. European Journal of Chemistry. 2015;**6**:254-260. DOI: 10.5155/eurjchem.6.3.254-260.1233

[69] El-Azab AS, Eltahir KE. Synthesis and anticonvulsant evaluation of some new 2,3,8-trisubstituted-4(3H)-quinazoline derivatives. Bioorganic & Medicinal Chemistry Letters. 2011;**22**:327-333. DOI: 10.1016/j.bmcl.2011.11.007

[70] Al-Salem HS, Hegazy GH, El-Taher KF, El-Messery SM, Al-Obaid AM, El-Subbagh HI. Synthesis, anticonvulsant activity and molecular modeling study of some new hydrazinecarbothioamide, benzenesulfonohydrazide, and phenacylacetohydrazide analogues of 4(3H)-quinazolinone. Bioorganic & Medicinal Chemistry Letters. 2015;**25**:1490-1499. DOI: 10.1016/j.bmcl.2015.02.025

[71] Popova E, Kostov M, Kupenova P. Effects of dopamine D1 receptor blockade on the ERG b- and d-waves during blockade of ionotropic GABA receptors. Eye and Vision (London). 2016;**3**:32. DOI: 10.1186/s40662-016-0064-4

[72] Abuelizz HA, Dib RE, Marzouk M, Anouar EH, A Maklad Y, N Attia H, Al-Salahi R. Molecular Docking and Anticonvulsant Activity of Newly Synthesized Quinazoline Derivatives. Molecules. 2017;**22**:1094. DOI: 10.3390/molecules22071094

[73] Jatav V, Mishra P, Kashaw S, Stables JP. CNS depressant and anticonvulsant activities of some novel 3-[5-substituted 1,3,4-thiadiazole-2-yl]-2-styryl quinazoline-4(3H)-ones. European Journal of Medicinal Chemistry. 2008;**43**:1945-1954. DOI: 10.1016/j.ejmech.2007.12.003

[74] Patel NB, Patel JC. Synthesis and antimicrobial activity of Schiff bases and 2-zetidinones derived from quinazolin-4(3H)-one. Arabian Journal of Chemistry. 2011;**4**:403-411

[75] Jantová S, Stankovsky S, Spirkova S. *In vitro* antibacterial activity of ten series of substituted quinazolines. Biologia Bratislava. 2004;**59**:741-752

[76] Vani KV, Ramesh G, Rao CV. Synthesis of new triazole and oxadiazole derivatives of quinazolin-4(3H)-one and their antimicrobial activity. Journal of Heterocyclic Chemistry. 2016;**53**:719-726. DOI: 10.1002/jhet.2353

[77] Kohli D, Hashim SR, Vishal S, Sharma M, Singh AK. Synthesis and

antibacterial activity of quinazolinones derivatives. International Journal of Pharmacy and Pharmaceutical Sciences. 2009;**1**:163-169. Available from: https://innovareacademics.in/journal/ijpps/Vol%201%20Issue%201/130.pdf [Accessed: 09 November 2019]

[78] Sowjanya C, Bharathi VR, Devi GK, Rajitha G. Synthesis and evaluation of some novel 3-[5-phenyl-1,3,4-oxadiazole-2-yl]-2-(substituted styryl)-quinazoline-4(3H)-ones for antibacterial activity. Journal of Chemical and Pharmaceutical Research. 2011;**3**:212-216. Available form: www.jocpr.com [Accessed: 09 September 2019]

[79] El-Hashash MA, Guirgui DB, El-Badry YA. Synthesis and evaluation of new 2, 3- and 2, 4-disubstituted quinazoline derivatives as potential antibacterial and antifungal agents. Der Pharma Chemica. 2011;**3**:147-159. Available from: http://derpharmachemica.com/archive.html [Accessed: 09 September 2019]

[80] Zhu S, Wang J, Chandrashekar G, Smith E, Liu X, Zhang Y. Synthesis and evaluation of 4-quinazolinone compounds as potential antimalarial agents. European Journal of Medicinal Chemistry. 2010;**45**:3864-3869. DOI: 10.1016/j.ejmech.2010.05.040

[81] Birhan YS, Bekhit AA, Hymete A. In vivo antimalarial evaluation of some 2,3-disubstituted-4(3H)-quinazolinone derivatives. BMC Research Notes. 2015;**8**:589. DOI: 10.1186/s13104-015-1578-x

[82] Wang Z, Wang M, Yao X, Li Y, Tan J, Wang L, et al. Design, synthesis and antiviral activity of novel quinazolinones. European Journal of Medicinal Chemistry. 2012;**53**:275-282. DOI: 10.1016/j.ejmech.2012.04.010

[83] Gao X, Cai X, Yan K, Song B, Gao L, Chen Z. Synthesis and antiviral bioactivities of 2-aryl- or 2-methyl-3-(substituted-benzalamino)-4(3H)-quinazolinone derivatives. Molecules. 2007;**12**:2621-2642. DOI: 10.3390/12122621

[84] Liu S, Wang W, Jiang L, Wan S, Zhang L, Yu R, et al. 2-Pyridinyl-4(3H)-quinazolinone: A scaffold for anti-influenza a virus compounds. Chemical Biology & Drug Design. 2015;**86**:1221-1225. DOI: 10.1111/cbdd.12589

[85] Carlos L-O, Lynn M. Emerging roles of proteases in tumour suppression. Nature Reviews. Cancer. 2007;**7**:800-808. DOI: 10.1038/nrc2228

[86] Singh M, Raghav N. 2,3-Dihydroquinazolin-4(1H)-one derivatives as potential non-peptidyl inhibitors of cathepsins B and H. Bioorganic Chemistry. 2015;**59**:12-22. DOI: 10.1016/j.bioorg.2015.01.005

[87] Raghav N, Singh M. Design, synthesis and docking studies of bischalcones based quinazoline-2(1H)-ones and quinazoline-2(1H)-thiones derivatives as novel inhibitors of cathepsin B and cathepsin H. European Journal of Pharmaceutical Sciences. 2014;**54**:28-39. DOI: 10.1016/j.ejps.2013.12.018

[88] Pommier Y. DNA topoisomerase I inhibitors: Chemistry, biology, and interfacial inhibition. Chemical Reviews. 2009;**109**:2894-2902. DOI: 10.1021/cr900097c

[89] Liang JL, Cha HC, Jahng Y. Recent advances in the studies on luotonins. Molecules. 2011;**16**:4861-4883. DOI: 10.3390/molecules16064861

[90] Ibric A, Eckerstorfer S, Eder M, Louko I, Tunjic L, Heffeter P, et al. Position-selective synthesis and biological evaluation of four isomeric A-ring amino derivatives of the alkaloid luotonin A. Molecules. 2019;**24**:716. DOI: 10.3390/molecules24040716

[91] Khadka DB, Tran GH, Shin S, Nguyen HTM, Cao HT, Zhao C, et al. Substituted 2-arylquinazolinones: Design, synthesis, and evaluation of cytotoxicity and inhibition of topoisomerases. European Journal of Medicinal Chemistry. 2015;**103**:69-79. DOI: 10.1016/j.ejmech.2015.08.040

[92] Kamata JJ, Okada T, Kotake YKN, Uenaka T, Yamaguchi A, Tsukahara K, et al. Synthesis and evaluation of novel pyrimidoacridone, –phenoxazine, and -carbazole as topoisomerase II inhibitors. Chemical & Pharmaceutical Bulletin (Tokyo). 2004;**52**:1071-1081. DOI: 10.1248/cpb.52.1071

[93] Saeedi M, Mohammadi-Khanaposhtani M, Pourrabia P, Razzaghi N, Ghadimi R, Imanparast S, et al. Design and synthesis of novel quinazolinone-1,2,3-triazole hybrids as new anti-diabetic agents: *In vitro* α-glucosidase inhibition, kinetic, and docking study. Bioorganic Chemistry. 2019;**83**:161-169. DOI: 10.1016/j.bioorg.2018.10.023

[94] Abuelizz HA, Anouar E, Ahmad R, Azman NIIN, Marzouk M, Al-Salah R. Triazoloquinazolines as a new class of potent α-glucosidase inhibitors: in vitro evaluation and docking study. PLoS One. 2019;**14**(8):e0220379. DOI: 10.1371/journal.pone.0220379

[95] Javaid K, Saad SM, Rasheed S, Moin ST, Syed N, Fatima I, et al. 2-Arylquinazolin-4(3H)-ones: A new class of α-glucosidase inhibitors. Bioorganic & Medicinal Chemistry. 2015;**23**:7417-7421. DOI: 10.1016/j.bmc.2015.10.038

[96] Rahman MU, Rathore A, Siddiqui AA, Parveen G, Yar MS. Synthesis and characterization of quinazolinederivatives: Search for hybrid molecule as diureticand antihypertensive agents. Journal of Enzyme Inhibition and Medicinal Chemistry. 2014;**29**:733-743. DOI: 10.3109/14756366.2013.845820

[97] Stroud RM, Santi DV, Hardy LW, Montfort WR, Jones MO, Finer-Moore JS. Atomic structure of thymidylate synthase: Target for rational drug design. Science. 1987;**235**:448-455. DOI: 10.1126/science.3099389

[98] Peters GJ, Backus HH, Freemantle S, van Triest B, Codacci-Pisanelli G, van der Wilt CL, et al. Induction of thymidylate synthase as a 5-fluorouracil resistance mechanism. Biochimica et Biophysica Acta. 2002;**1587**(2-3):194-205. DOI: 10.1016/S0925-4439(02)00082-0

[99] Wang M, Yang J, Yuan M, Xue L, Li H, Tian C, et al. Synthesis and antiproliferative activity of a series of novel 6-substituted pyrido[3,2-d]pyrimidines as potential nonclassical lipophilic antifolates targeting dihydrofolate reductase. European Journal of Medicinal Chemistry. 2017;**128**:88-97. DOI: 10.1016/j.ejmech.2017.01.033

[100] El-Messery SM, Hassan GS, Nagi MN, Habib ESE, Al-Rashood ST, El-Subbagh HI. Synthesis, biological evaluation and molecular modeling study of some new methoxylated 2-benzylthio-quinazoline-4(3H)-ones as nonclassical antifolates. Bioorganic & Medicinal Chemistry Letters. 2016;**26**:4815-4823. DOI: 10.1016/j.bmcl.2016.08.022

[101] El-Gazzar YI, Georgey HH, El-Messery SM, Ewida HA, Hassan GS, Raafat MM, et al. Synthesis, biological evaluation and molecular modeling study of new (1,2,4-triazole or 1,3,4-thiadiazole)-methylthio-derivatives of quinazolin-4(3H)-one as DHFR inhibitors. Bioorganic Chemistry. 2017;**72**:282-292. DOI: 10.1016/j.bioorg.2017.04.019

[102] Javaid S, Saad SM, Perveen S, Khan KM, Choudhary MI. 2-Arylquinazolin-4(3H)-ones: A novel class of thymidine phosphorylase inhibitors. Bioorganic Chemistry. 2015;**63**:142-151. DOI: 10.1016/j.bioorg.2015.10.006

[103] Lum CT, Stahl SM. Opportunities for reversible inhibitors of monoamine oxidase–A (RIMAs) in the treatment of depression. CNS Spectrums. 2012;**17**(3):107-120. DOI: 10.1017/s1092852912000594

[104] Fernandez HH, Chen JJ. Monoamine oxidase-B inhibition in the treatment of Parkinson's disease. Pharmacotherapy. 2007;**27**(12 Pt 2):174S-185S. DOI: 10.1592/phco.27.12part2.174S

[105] Mostert S, Petzer A, Petzer JP. Indanones as high-potency reversible inhibitors of monoamine oxidase. ChemMedChem. 2015;**10**(5):862-873. DOI: 10.1002/cmdc.201500059

[106] Qhobosheane MA, Petzer A, Petzer JP, Legoabe LJ. Synthesis and evaluation of 2-substituted 4(3H)-quinazolinone thioether derivatives as monoamine oxidase inhibitors. Bioorganic & Medicinal Chemistry. 2018;**26**:5531-5537. DOI: 10.1016/j.bmc.2018.09.032

[107] Khattab SN, Haiba NS, Asal AM, Bekhit AA, Amer A, Abdel-Rahman HM, et al. Synthesis and evaluation of quinazoline amino acid derivatives as mono amine oxidase (MAO) inhibitors. Bioorganic & Medicinal Chemistry. 2015;**23**:3574-3585. DOI: 10.1016/j.bmc.2015.04.021

[108] Hughes AN, Rafi I, Griffin MJ. Phase I studies with the nonclassical antifolate nolatrexed dihydrochloride (AG337, THYMITAQ) administered orally for 5 days. Clinical Cancer Research. 1999;5(1):111-118. PMID: 9918208

[109] Cohen E, Klarberg B, Vaughan JR. Quinazolinone sulfonamides. A new class of diuretic agents. Journal of the American Chemical Society. 1960;**82**:2731-2735. DOI: 10.1021/ja01496a020

[110] Maggi GC, Donati C, Gueli AD. Single dose pharmacokinetics of fenquizone in healthy volunteers. Arzneimittel-Forschung. 1985;**35**:994-998. PMID: 4026929

[111] Guedes PM, Urbina JA, de Lana M, Afonso LC, Veloso VM, Tafuri WL, et al. Activity of the new triazole derivative albaconazole against Trypanosoma (Schizotrypanum) cruzi in dog hosts. Antimicrobial Agents and Chemotherapy. 2004;**48**(11):4286-4292. DOI: 10.1128/AAC.48.11.4286-4292.2004

[112] McLaughlin NP, Evans P. Dihydroxylation of vinyl sulfones: Stereoselective synthesis of (+)- and (−)-febrifugine and halofuginone. The Journal of Organic Chemistry. 2010;**75**(2):518-521. DOI: 10.1021/jo902396m

[113] Ochiai T, Ishida R. Pharmacological studies on 6-amino-2-fluoromethyl-3-(O-tolyl)-4(3H)-quinazolinone (afloqualone), a new centrally acting muscle relaxant. (II) Effects on the spinal reflex potential and the rigidity. Japanese Journal of Pharmacology. 1982;**32**(3):427-438. DOI: 10.1254/jjp.32.427

[114] Kobayashi Y. Capsaicin-like anti-obese activities of evodiamine from fruits of *Evodia rutaecarpa*, a vanilloid receptor agonist. Planta Medica. 2001;**67**(7):628-633. DOI: 10.1055/s-2001-17353

Section 2

Synthesis and Biological Activity

Chapter 3

Synthesis of Quinazoline and Quinazolinone Derivatives

Heba E. Hashem

Abstract

Active heterocyclic compounds are one of the main topics of interest for the medicinal chemists as they display a number of pharmacological activities. Nitrogen, sulfur, and oxygen containing five- and six-membered heterocyclic compounds have occupied enormous significance in the field of medicinal chemistry. The most important six-membered heterocyclic compounds are quinazoline and quinazolinone derivatives for their biological activities. The current chapter outlined the different methods for synthesis of quinazoline and quinazolinone derivatives that possess broad spectrum of biological activities.

Keywords: quinazoline, quinazolinone synthesis, six-membered heterocycles, biological activity

1. Introduction

Quinazoline (1,3-diazanaphthalene or 5,6-benzopyrimidine) and 4(3H)-quinazolinone derivatives have a great interest in organic synthesis and medicinal chemistry fields as they possess a broad range of pharmacological activities. They exhibit antimicrobial [1], antimalarial [2], antioxidant [3], anti-inflammatory [4], anticonvulsant [5], antihypertensive [6], antidiabetic [7], and antitumor activities [8–10].

Many quinazolinone derivatives occurred naturally in various classes of the plant kingdom, microorganisms, and different animals (**Figure 1**). The first discovery of quinazolinone alkaloid is ***febrifugine*** which possesses antimalarial potential, extracted from the Chinese plant ***aseru*** (*Dichroa febrifuga* Lour) [11].

Quinazoline is a heterocyclic compound of two fused six-membered simple aromatic rings—benzene and pyrimidine ring. It is a yellow-colored compound, found usually in crystalline form. Its oxo-derivative (quinazolinone) is classified into three types according to the position and number of carbonyl group: 2(1H) quinazolinones, 4(3H)quinazolinones, and 2,4(1H,3H)quinazolinedione (**Figure 2**).

2. Chemistry of quinazoline

Quinazoline is a compound made up of two fused six-membered simple aromatic rings—benzene and pyrimidine ring. The properties of the pyrimidine ring were affected by the presence of fused benzene ring. The two nitrogen atoms are not equivalent, and the marked polarization of the 3,4-double bond is reflected

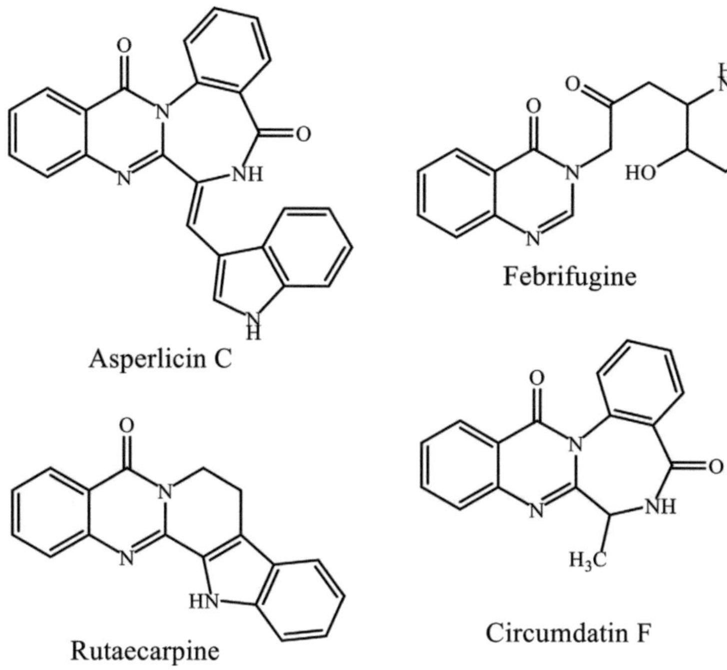

Figure 1.
Structure of different quinazolinone alkaloids.

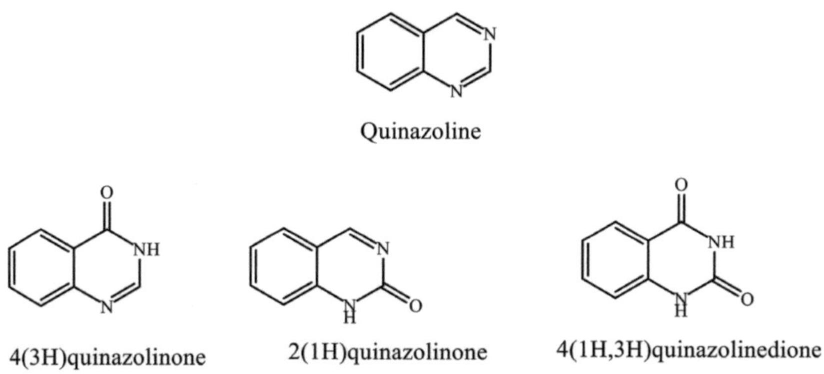

Figure 2.
Structure of quinazoline and different quinazolinone compounds.

in the reactions of quinazoline. The properties of quinazoline derivatives depend on the following three factors:

a. The nature of the substituents

b. The presence of substituent whether they are in the pyrimidine ring or in the benzene ring

c. The presence of conjugation in the pyrimidine ring

The first synthesized quinazoline in laboratory was achieved by Gabriel in 1903 [12]. Most of quinazoline derivatives are stable in cold acidic or basic medium but can be destroyed at high temperature and undergo ring opening reaction, affording

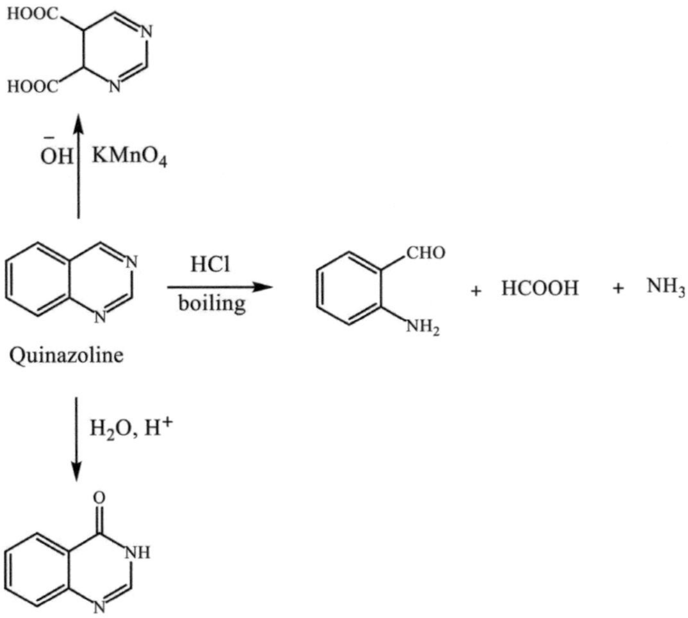

Figure 3.
Oxidation reaction of quinazoline at different medium.

O-aminobenzaldehyde, ammonia, and formic acid. Quinazoline derivative can be easily oxidized in acidic medium at room temperature to give 3,4-dihydro-4-oxo quinazoline, while in alkaline medium using potassium permanganate will afford 3,4-dihydro-6 4-oxo quinazoline (cf. **Figure 3**).

3. Spectral characterization of quinazoline and quinazolinone derivatives

The spectroscopic analysis of some synthesized quinazoline and quinazolinone derivatives was studied to investigate their structures including infrared, mass spectroscopy, ^{1}HNMR, and elemental analysis. The resulted data could be taken as standard for the new synthesized quinazoline analogue [13].

3.1 Infrared spectra

Quinazoline derivatives found to give mainly three absorption bands in IR spectra: 1478–1517, 1566–1581, and 1612–1628 cm^{-1}; these represented bands are correlated to C–N, C=C, and C=N groups, while quinazolinone compounds showed 1680–1700 and 1640–1660 cm^{-1} corresponding to C=O and C=N groups [13, 14].

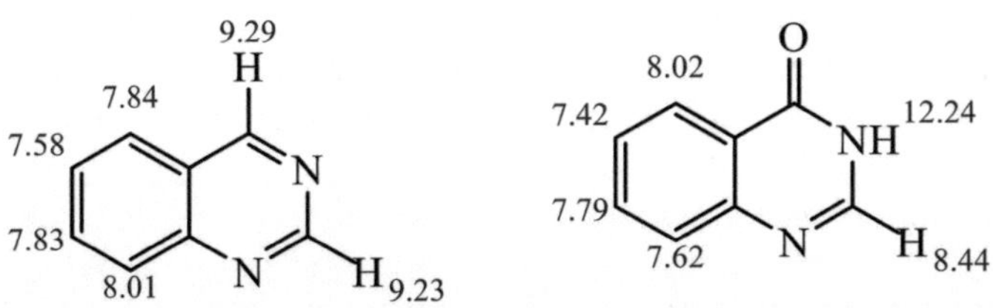

(I) (II)

The ^{1}HNMR spectra of quinazoline and quinazolinone derivatives are different from each other according to the presence of acidic proton and its position in the presented compound. In general the 1HNMR spectrum of the main quinazoline (I) represents multiple signals in the aromatic region δ 7–8 and two singlet signals for the two CH=N protons at δ 9–9.5 ppm, while quinazolinone (II) will show also signals of aromatic protons in the same region as well as one singlet signal for CH=N proton and one broad singlet signal at the down-field region for the NH proton at δ 12–13 ppm [13, 14].

On the other hand, the ^{13}C NMR spectrum for quinazoline and quinazolinone derivatives is nearly the same, as it shows signals at δ 100–160 ppm region.

4. Synthesis of quinazoline and quinazolinone derivatives

The synthesis of various quinazoline compounds is largely based on the substitution patterns of the 1,3-diazine moiety of the system. The first quinazoline derivative (2-cyano-3,4-dihydro-4-oxoquinazoline) was synthesized in 1869 by the reaction of cyanogens with anthranilic acid [15]. Many years later quinazoline was obtained by decarboxylation of the 2-carboxy derivative (quinazolinone) which can be synthesized more easily by a different method.

4.1 Synthesis of quinazolinone

4.1.1 Niementowski's synthesis

From anthranilic acid and formamide.

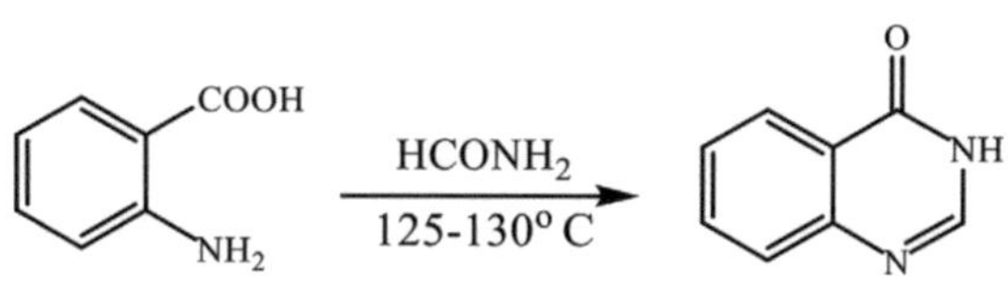

4.1.2 Grimmel, Guinther, and Morgan's synthesis

From the reaction of o-amino benzoic acid with amine in the presence of phosphorus trichloride in toluene.

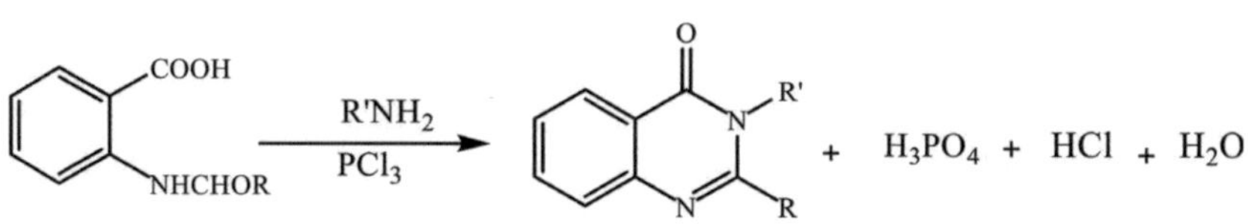

4.1.3 From 3,1,4-benoxazones (acylanthranils) and amines

Various quinazoline and quinazolinone derivatives can be synthesized from the reaction of benzoxazinone and different amine compounds in different media.

4.1.3.1 Reaction with ammonium hydroxide

When ammonium hydroxide reacted with benzoxazinone (1) over 1–3 h, it produced anthranilamides (2) which cyclizes to 4-quinazolones (3) under thermal conditions (240–280°C) or on heating with acetic anhydride [16, 17].

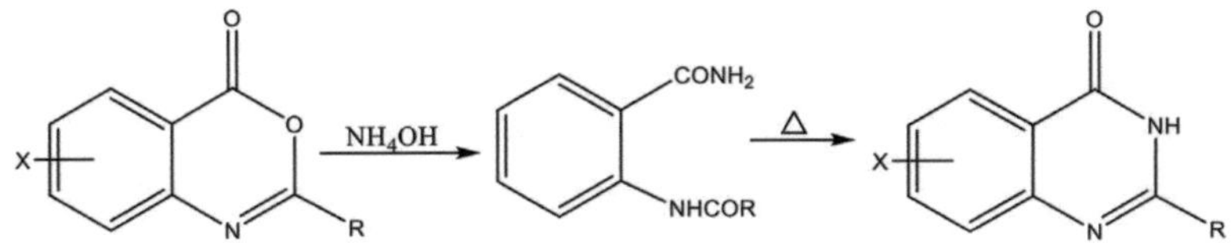

(1) (2) (3)

X= H, Cl, Br, Me
R= alkyl, PhY(Y=Br, NHMe), CH_2CN

4.1.3.2 Reaction with different aromatic amines

It was stated by several authors that 2-substituted benzoxazinone reacted easily with primary aromatic amines, giving the corresponding quinazolones (4) [18].

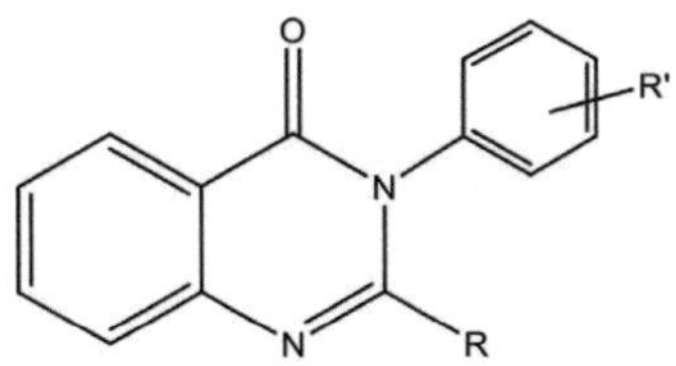

(4)

On the other hand, reaction of benzoxazinone (5) with o-phenylenediamine gave quinazolinone derivative (6) or the fused quinazoline derivative (7) according to the reaction medium [19].

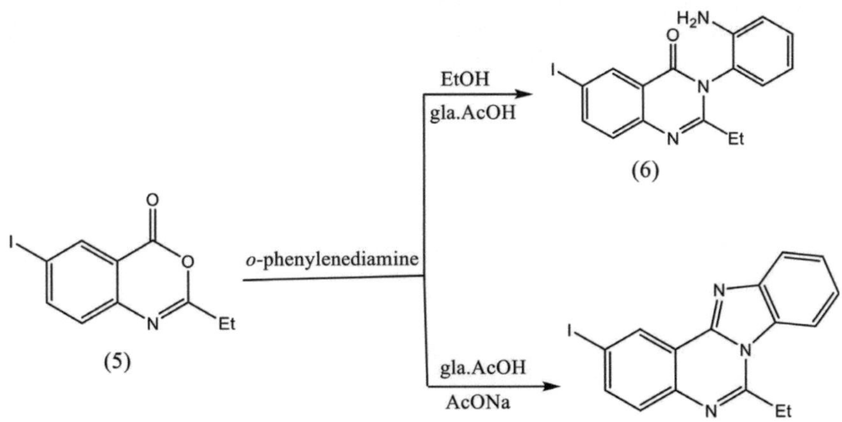

(5) (6) (7)

4.1.3.3 Reaction with hydrazine hydrate

It was reported that benzoxazinone (8) reacted with hydrazine hydrate in ethanol and has the corresponding quinazolinone (9), while carrying out the same reaction in boiling acetic acid glacial afforded the fused quinazoline (10) [13].

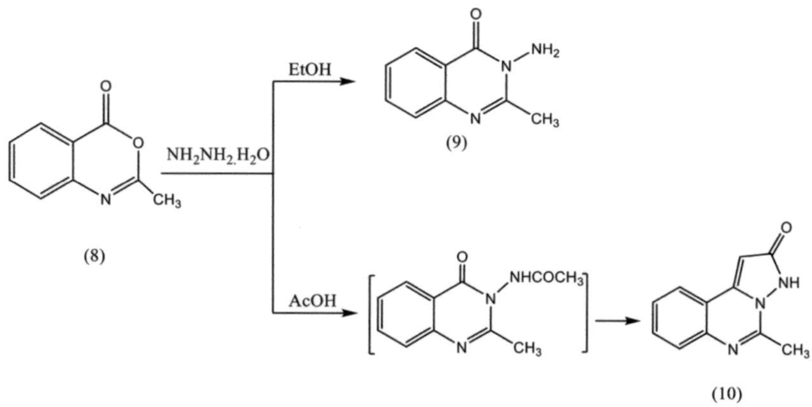

4.1.3.4 Reaction with different carbohydrazide

Treatment of 2-substituted-3,1-benzoxazin-4-ones (11) with semicarbazide hydrochloride in dry pyridine is a good way to construct a third heterocyclic ring condensed with quinazoline (12) [18].

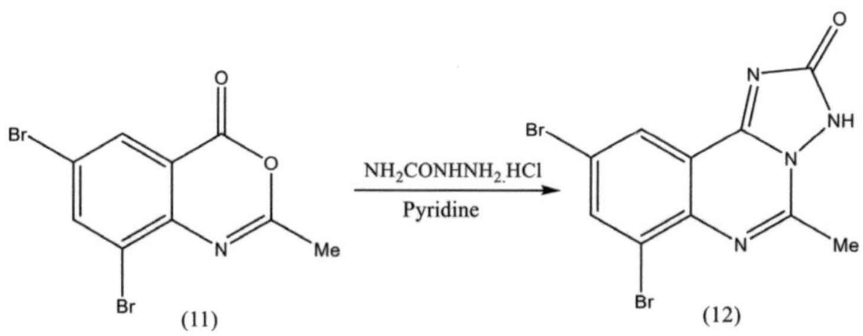

The reaction of benzooxazinone (8) with 2-benzamido-3-phenylacrylohydrazide (13) glacial acetic acid in the presence of fused sodium acetate gave quinazoline derivative (14). In contrast, their reaction in pyridine afforded pyrazoloquinazoline derivative (15) [13].

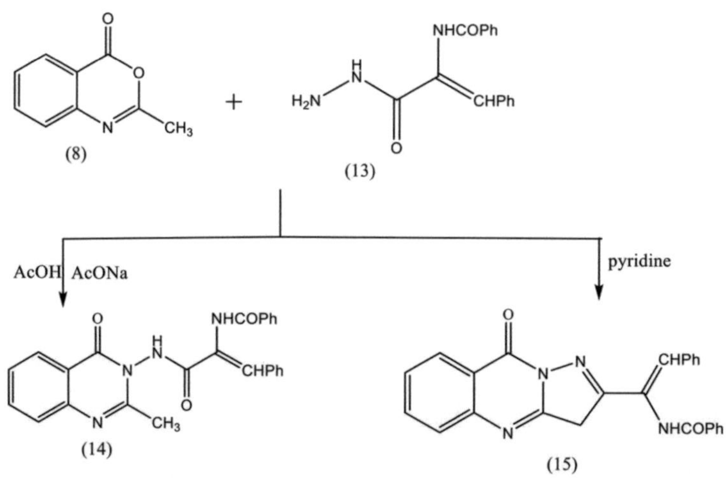

Reaction of benzoxazinone (8) with cyanoacetohydrazide gave the corresponding cyano quinazolinone (16) which was reacted with different nucleophiles to give fused quinazoline and annulated quinazolinone derivatives (17–19) [13].

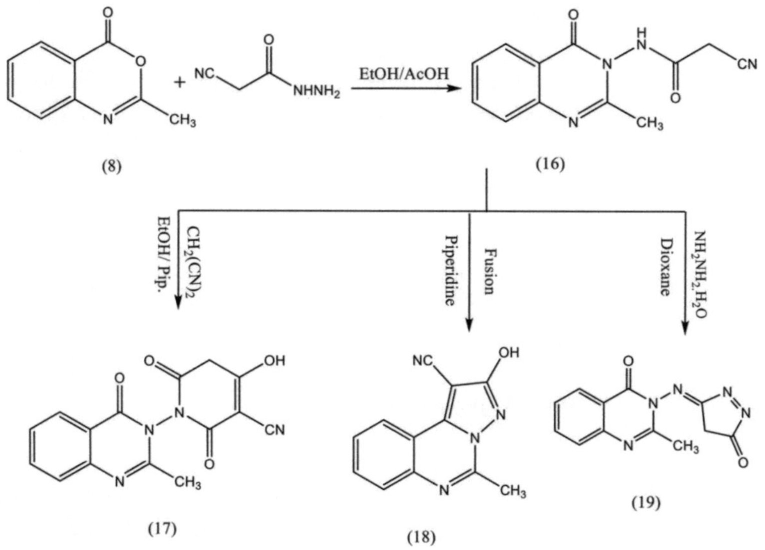

It was also reported that refluxing an equimolar amount of the benzoxazinone (8) with thiocarbonohydrazide in ethanol and in the presence of few drops of glacial acetic acid furnished quinazolinone (20) in the two isomers of thione and thiol form [13].

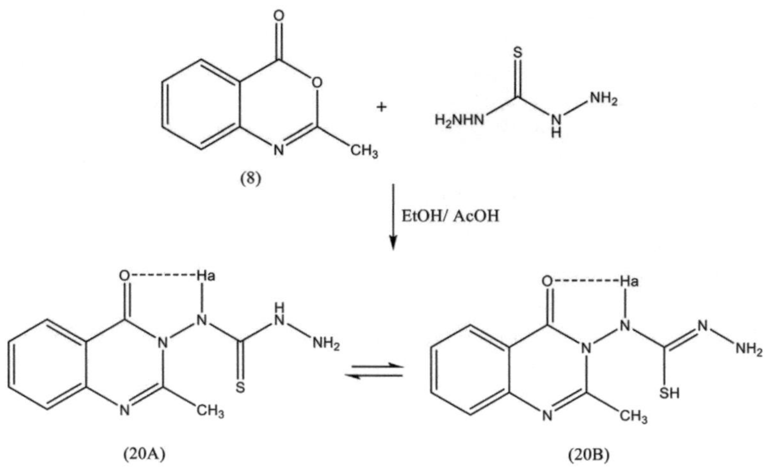

4.1.4 Sen and Ray's synthesis

Isobutyrylanilides with urethane and phosphorus pentoxide in xylene gave 2-propyl- and 2-isopropyl-3,4-dihydro-4-oxoquinazolines.

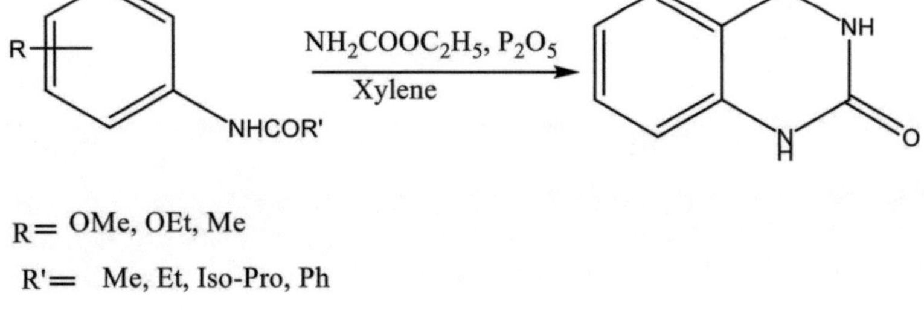

4.1.5 From 2-aminobenzylamine

Reaction of 2-aminobenzylamine with butyrolactone further condensed with benzaldehyde afforded 3-(2-chlorobenzylidene)-1,2,3,9-tetrahydropyrrolo-2-quinazoline.

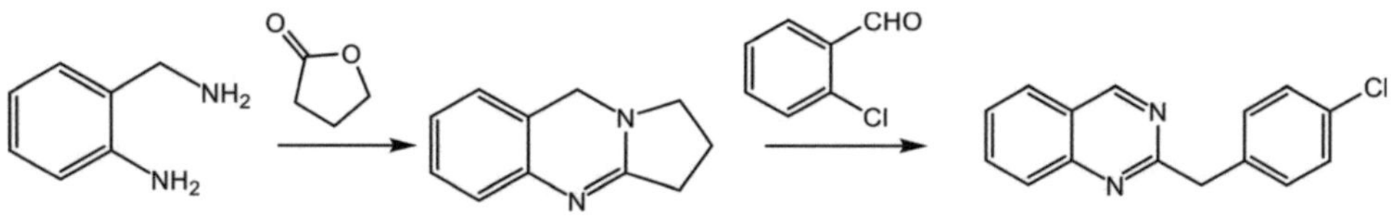

5. Biological importance of quinazoline derivatives

As we mentioned above, the important biological activity of quinazoline and quinazolinone skeletons in various fields depends mainly on the substituents of quinazoline compounds. Different substituted quinazoline compounds are found to be active as antihypertensive, antineoplastic, antidepressant, and antipsychotic, and others are effective against analgesic, antipsychotic, antiarrhythmic, cancer, and other activities [20–22].

5.1 Anticancer

It was reported that 3-substituted quinazolin-4(3H)-ones and 3,4-dihydroquinazolin-2-(1H)-one derivatives possess broad spectrum antitumor activities toward different cell (**Figure 4**) [23].

Also, different quinazoline derivatives containing thiosemicarbazide moiety possess antitumor activity (**Figure 5**) [24].

5.2 Antibacterial activity

It was reported that some novel substituted iodoquinazoline derivatives possess remarkable activity toward Gram-negative bacteria *E. coli* (**Figure 6**) [25].

5.3 Antiviral agents

A series of Schiff bases of some 2-phenyl quinazoline-4(3)H-one derivatives have shown great activity as antiviral agents (**Figure 7**) [26].

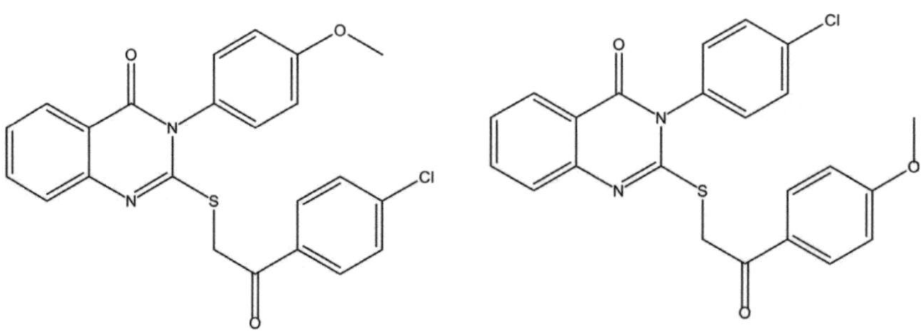

Figure 4.
Anti-tumor quinazolinone derivatives.

DOI: http://dx.doi.org/10.5772/intechopen.89180

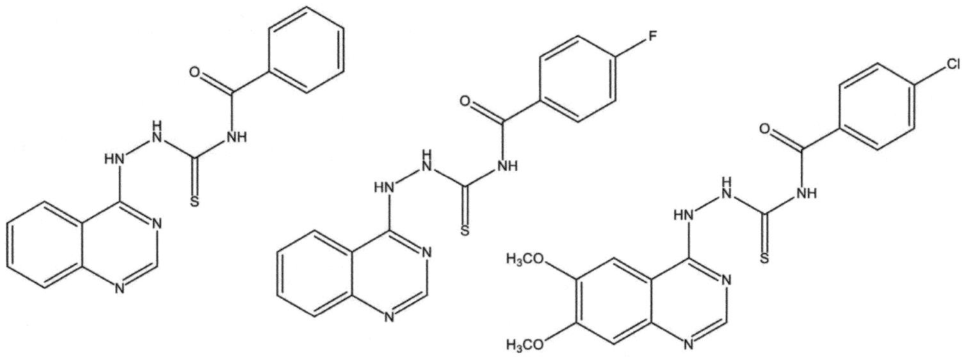

Figure 5.
Quinazoline derivatives bearing thiosemicarbazide possess anti-tumor activity.

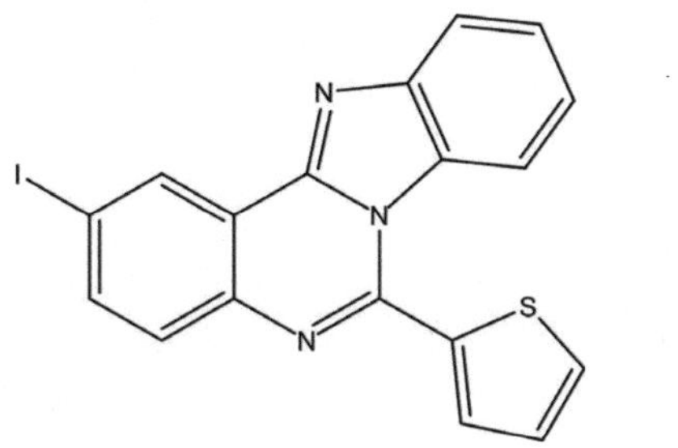

Figure 6.
Quinazoline derivatives with antibacterial activity.

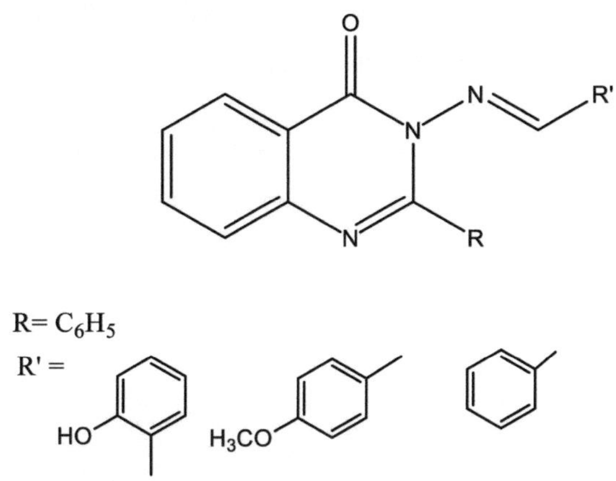

Figure 7.
Different schiff base of quinazolinone with antiviral activity.

5.4 Antimutagenic activity

The (S)-4-aminoquinazoline alcohols performed great antimutagenic activity when tested by using *Salmonella typhimurium* and *E. coli* WP2uvrA tester strains at 0.01, 0.1, and 1 lg/plate concentrations (**Figure 8**) [27].

5.5 Antioxidant activity

Some novel thiazoloquinazoline derivatives are investigated for antioxidant activity by DPPH radical assay, nitric oxide scavenging activity, and hydrogen

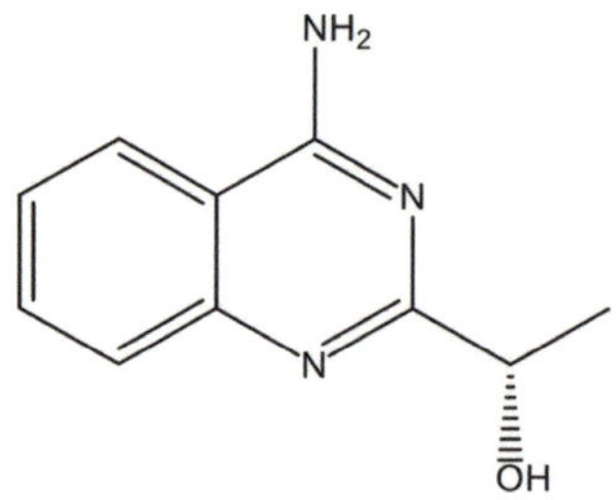

Figure 8.
Antimutagenic activity of amino quinazoline derivative.

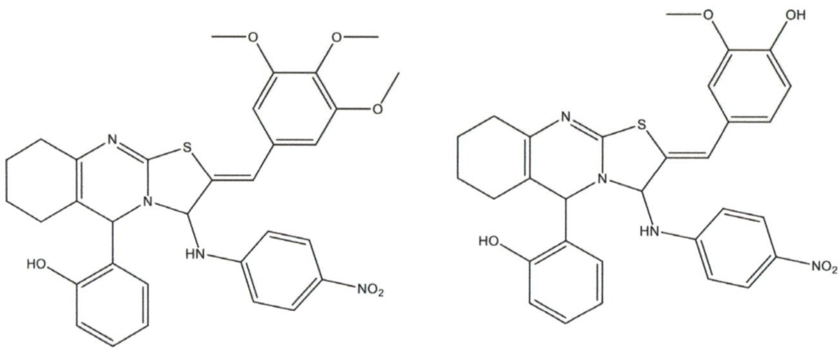

Figure 9.
Antioxidant activity of different quinazoline.

peroxide scavenging activity and possess high potent antioxidant activity (**Figure 9**) [28].

6. Conclusion

Quinazoline and quinazolinone compounds which have a lot of considerable pharmacological interests can be synthesized by different methods, and the most attractive method was carried out starting from benzoxazinone derivatives.

Author details

Heba E. Hashem
Department of Chemistry, College for Girls, Ain Shams University, Heliopolis, Cairo, Egypt

*Address all correspondence to: hebahashem89@yahoo.com

References

[1] Grover G, Kini SG. Synthesis and evaluation of new quinazolone derivatives of nalidixic acid as potential antibacterial and antifungal agents. European Journal of Medicinal Chemistry. 2006;**41**:256-262

[2] Verhaeghe P, Azas N, Gasquet M, Hutter S, Ducros C, Laget M, et al. Synthesis and antiplasmodial activity of new 4-aryl-2-trichloromethyl quinazolines. Bioorganic & Medicinal Chemistry Letters. 2008;**18**:396-401

[3] Roopan SM, Maiyalagan T, Khan FN. Solvent-free syntheses of some quinazolin-4(3H)-ones derivatives. Canadian Journal of Chemistry. 2008; **86**:1019

[4] Smits RA, Adami M, Istyastono EP, Zuiderveld OP, van Dam CME, de Kanter FJJ, et al. Synthesis and QSAR of quinazoline sulfonamides as highly potent human histamine H4receptor inverse agonists. Journal of Medicinal Chemistry. 2010;**53**:2390-2400

[5] Georgey H, Abdel Gawad N, Abbas S. Synthesis and anticonvulsant activity of some quinazolin-4-(3H)-one derivatives. Molecules. 2008;**13**: 2557-2569

[6] Jain KS, Bariwal JB, Kathiravan MK, Phoujdar MS, Sahne RS, Chauhan AK, et al. Recent advances in selective alpha1-adrenoreceptor B.S. antagonists as antihypertensive agents. Bioorganic & Medicinal Chemistry. 2008;**16**: 4759-4800

[7] Malamas MS, Millen J. Quinazoline acetic acids and related analogues as aldose reductase inhibitors. Journal of Medicinal Chemistry. 1991;**34**: 1492-1503

[8] Shallal HM, Russu WA. Discovery, synthesis, and investigation of the antitumor activity of novel piperazinylpyrimidine derivatives. European Journal of Medicinal Chemistry. 2011;**46**:2043-2057

[9] Chilin A, Conconi MT, Marzaro G, Guiotto A, Urbani L, Tonus F, et al. Exploring epidermal growth factor receptor (EGFR) inhibitor features: The role of fused dioxygenated rings on the quinazoline scaffold. Journal of Medicinal Chemistry. 2010;**53**: 1862-1866

[10] Sagiv-Barfi I, Weiss E, Levitzki A. Design, synthesis, and evaluation of quinazoline T cell proliferation inhibitors. Bioorganic & Medicinal Chemistry. 2010;**18**:6404-6413

[11] Wattanapiromsakul C, Forster PI, Waterman PG. Alkaloids and limonoids from *Bouchardatia neurococca*: Systematic significance. Phytochemistry. 2003;**64**:609-615

[12] Gabriel S. Synthesis and pharmacological evaluation of 3-alkyl/ aryl–2-methylquinazolin-4-one derivatives. Bernoulli Society. 1903;**36**: 800-845

[13] Hemdan MM, Youssef ASA, El-Mariah FA, Hashem HE. Synthesis and antimicrobial assessments of some quinazolines and their annulated systems. Journal of Chemical Research. 2017;**41**:106-111

[14] Zeinab F, Nasrin R, Razieh S, Kamiar Z, Mohammad A, Soghra K. Synthesis of some novel dibromo-2-arylquinazolinone derivatives as cytotoxic agents. Research in Pharmaceutical Sciences. 2019;**14**(2): 115-121

[15] Armarego WLF. A Text Book of Quinazolines. 1963. pp. 1-320

[16] Essawy A, El-Hashash MA, El-Gendy AM, Hamed MMM. Synthesis

and studies on heterocyclic compounds containing mixed and non-mixed systems. Indian Journal of Chemistry. 1982;**21B**:593

[17] Webber SE, Bleckman TM, Attard J, Deal JG, Kathardekar V, Welsh KM, et al. Design of thymidylate synthase inhibitors using protein crystal structures: The synthesis and biological evaluation of a novel class of 5-substituted quinazolinones. Journal of Medicinal Chemistry. 1993;**36**:733-734

[18] Madkour HMF. Reactivity of 4H-3,1-benzoxazin-4-ones towards nitrogen and carbon nucleophilic reagents: Applications to the synthesis of new heterocycles. ARKIVOC. 2004;**36**

[19] Ahmed MA, El-Azab AS, Mohamed MA, Bakhat MA, Abdel-Hamid SG. Synthesis of some new substituted iodo quinazoline derivatives and their antimicrobial screening. Journal of Saudi Chemical Society. 2011; **15**:319-325

[20] Rajput R, Mishra AP. A review on biological activity of quinazolinones. International Journal of Pharmacy and Pharmaceutical Sciences. 2012;**4**(2): 66-70

[21] Pati B, Banerjee S. Quinazolines: An illustrated review. Journal of Advanced Pharmacy Education and Research. 2013;**3**(3):136-151

[22] Vijayakumar B, Prasanthi P, Teja KM. Quinazoline derivatives and pharmacological activities: A review. International Journal of Medicinal Chemistry & Analysis. 2013;**3**(1):10-21

[23] Abdel Gawad NM, Georgey HH, Youssef RM, El-Sayed NA. Synthesis and antitumor activity of some 2,3-disubstituted quinazolin-4(3H)-ones and 4,6-disubstituted-1,2, 3,4-tetrahydroquinazolin-2H-ones. European Journal of Medicinal Chemistry. 2010;**45**(12):6058-6067

[24] He J, Wang X, Zhao X, Liang Y, He H, Fu L. Synthesis and antitumor activity of novel quinazoline derivatives containing thiosemicarbazide moiety. European Journal of Medicinal Chemistry. 2012;**54**:925-930

[25] Alafeefy AM, El-Azab AS, Mohamed MA, Bakhat MA, Abdel-Hamid SG. Synthesis of some new substituted iodoquinazoline derivatives and their antimicrobial screening. Journal of Saudi Chemical Society. 2011; **15**(4):319-325

[26] Kumar KS, Ganguly S, Veerasamy R, DeClercq E. Synthesis, antiviral activity and cytotoxicity evaluation of Schiff bases of some 2-phenyl quinazoline-4(3)H-ones. European Journal of Medicinal Chemistry. 2010;**45**(11):5474-5479

[27] Kohli D, Hashim SR, Vishal S, Sharma M, Singh AK. Synthesis and antibacterial activity of quinazolinone derivatives. International Journal of Pharmacy and Pharmaceutical Sciences. 2009;**1**(1):163-169

[28] Selvam TP, Kumar PV, Kumar AS. Synthesis and anti-oxidant activity of novel 6,7,8,9 tetra hydro-5H-5-(2'-hydroxy phenyl)-2-(4'-substituted benzylidine)- 3-(4-nitrophenyl amino) thiazolo quinazoline derivatives. Research in Biotechnology. 2010;**1**(1): 38-48

Chapter 4

4(3*H*)-Quinazolinone Derivatives: Syntheses, Physical Properties, Chemical Reaction, and Biological Properties

Samir Y. Abbas

Abstract

4(3*H*)-Quinazolinone derivatives have considerable great interesting due to the diverse range of their biological properties. This review summarized the methods of preparation of 2-substituted-4(3*H*)-quinazolinone, 3-substituted-4(3*H*)-quinazolinone and 2,3-disubstituted-4(3*H*)-quinazolinone derivatives. Chemical reaction of 4(3*H*)-quinazolinone derivatives and the reactivity of the 2-methyl group, reactivity of the 3-amino group, electrophilic substitution, oxidation, reduction, reaction of 4(3*H*)-quinazolinones with metal ions, Mannich reaction, cycloaddition reaction as well as other reagents were discussed. Also, biological properties of 4(3*H*)-quinazolinone derivatives were given herein.

Keywords: 4(3*H*)-quinazolinones, quinazolines, syntheses, reaction, biological activity

1. Introduction

Quinazolinones are very significant heterocyclic compounds because of their potential pharmaceutical and biological activities. Quinazolinone derivatives reveal various medicinal properties such as analgesic, anti-inflammatory and anticancer activities, as well as antimicrobial activity. These heterocycles are valuable intermediates in organic synthesis. Therefore, various procedures have been developed for preparing these important compounds [1–5].

2. Syntheses of 4(3*H*)-quinazolinones

The syntheses of quinazolinones will be classified into the following three categories based on the substitution patterns of the ring system:

- 2-Substituted-4(3*H*)-quinazolinones
- 3-Substituted-4(3*H*)-quinazolinones
- 2,3-Disubstituted-4(3*H*)-quinazolinones

2.1 2-Substituted-4(3*H*)-quinazolinones

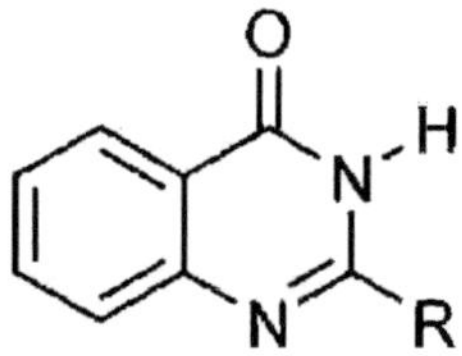

2.1.1 Amidation and cyclization of 2-aminobenzoic acid derivatives (anthranilic acid derivatives)

The most common approach involves amidation of 2-aminobenzoic acid derivatives (**Figure 1**). As an example, anthranilic acid derivatives **1** were coupled with the appropriate acid chloride to generate the corresponding substituted anthranilates **2** which underwent cyclization by treatment with acetic anhydride under reflux afforded the benzoxazin-4-ones **3**. Treatment of the benzoxazinones **3** with ammonia solution afforded the quinazolinone derivatives **4** [6]. Benzoxazinone derivatives can be obtained by treatment of anthranilic acid derivatives with acetic anhydride [7]. Also, the condensation of anthranilic acid derivatives with the *ortho* esters and ammonium acetate afforded the 2-substituted-4(3*H*)-quinazolinone derivatives [8, 9]. 2-Carboethoxy-quinazoline-4(3*H*)-one has been synthesized from the reaction of anthranilamide and diethyl oxalate [10].

2.1.2 Condensation of imidates with 2-aminobenzoic acid

Reaction between imidates and anthranilic acid **1** was reported for preparation of a series of quinazoline antifolate thymidylate synthase inhibitors [11]. The condensation of **1** and imidates **5** in methanol at 80°C afforded the desired quinazolinones **6** in good yield (**Figure 2**). The condensation reaction afforded the quinazolinones **6** in satisfactory to good yields. Connolly and Guiry extended this methodology to synthesize a series of 2-aryl- and 2-alkylquinazolinones [12].

2.1.3 Synthesis of quinazolinones from resin-bound isothioureas

A concise and efficient solid-phase synthesis of 2-amino-4(3*H*)-quinazolinones **9** has been reported by involving the reaction of polymer-bound isothioureas **8** with isatoic anhydride derivative **7** with good yield and purity (**Figure 3**) [13].

2.1.4 Hetero-Diels: alder synthesis of 2-substituted quinazolinones

Synthesis of 2-substituted-quinazolinones **12** was reported by the cyclisation of 1-aryl-4-dimethylamino-2-phenyl-1,3-diaza-1,3-butadienes **10** and phenyl

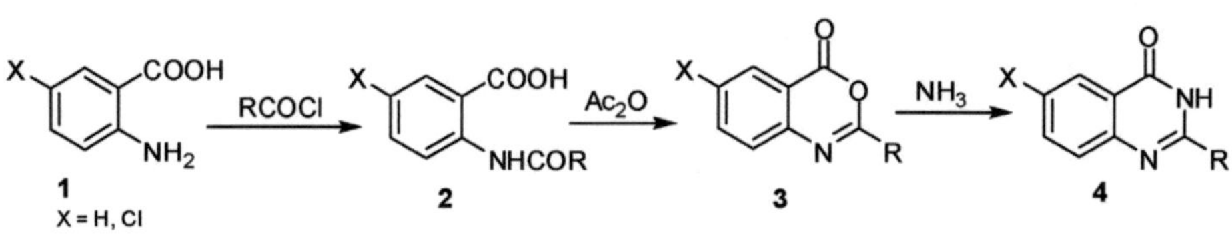

Figure 1.
Synthesis of 2-substituted-4(3H)-quinazolinones 4.

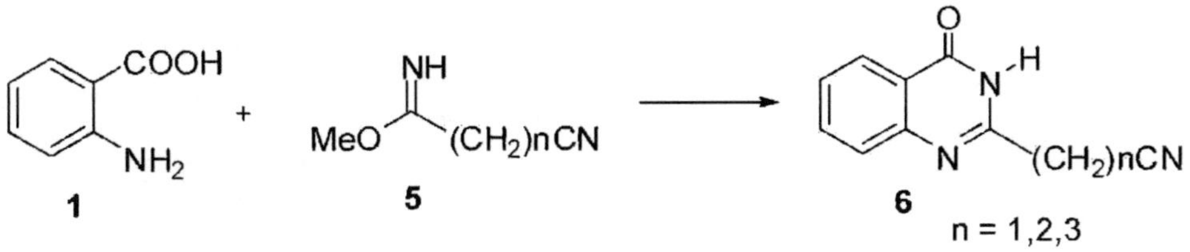

Figure 2.
*Synthesis of 2-substituted-4(3H)-quinazolinones **6**.*

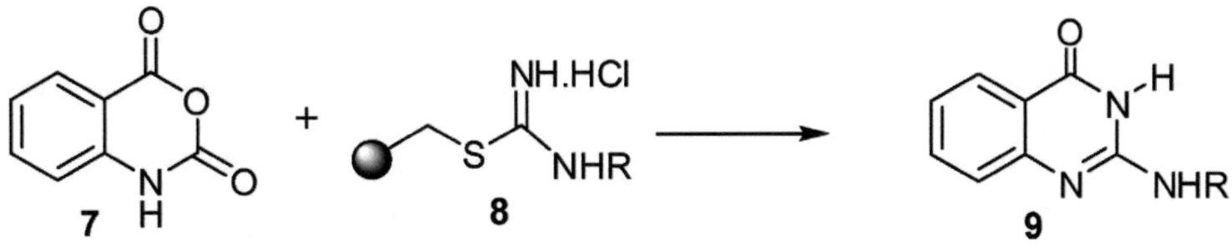

Figure 3.
*Synthesis of 2-amino-4(3H)-quinazolinones **9**.*

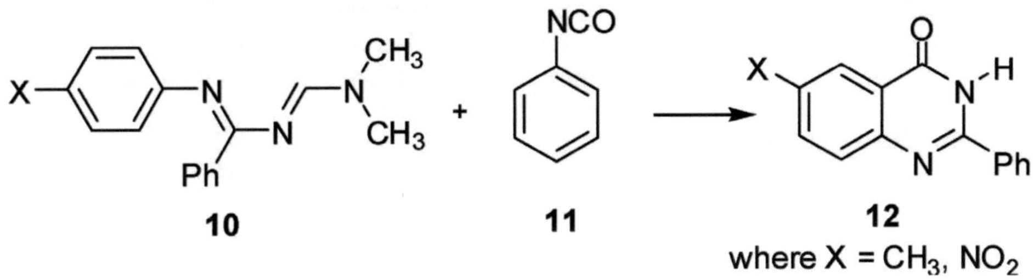

Figure 4.
*Synthesis of 2-substituted-4(3H)-quinazolinones **12**.*

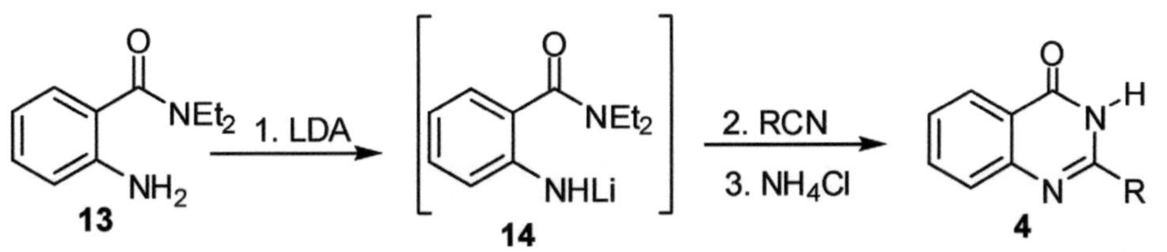

Figure 5.
*Synthesis of 2-substituted-4(3H)-quinazolinones **4**.*

isocyanate **11** [14]. The reaction was carried out under an atmosphere of nitrogen in toluene at reflux temperature, to furnish the desired products **12** in good yield (**Figure 4**).

2.1.5 Reaction of nitriles with lithiated 2-aminobenz- amides

A developed procedure for the synthesis of 2-aryl- and 2-alkyl-4(3*H*)-quinazolinones **4** by reaction of lithium 2-(diethyl aminocarbonyl) anilide **14** with the appropriate aryl or aliphatic nitrile has been reported [15]. This route is highly efficient when aryl and hetero-aryl nitriles were used. The intermediate **14** was prepared in situ by treating **13** with LDA in THF at −30°C (**Figure 5**).

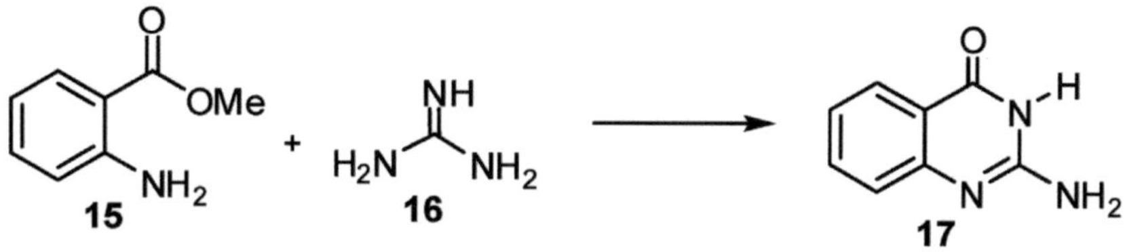

Figure 6.
Synthesis of 2-amino-4(3H)-quinazolinone 17.

2.1.6 Condensation of anthranilate esters with guanidine

The 2-amino-4(3*H*)-quinazolinone **17** has been prepared from the reaction of methyl anthranilate **15** with excess guanidine **16** in the presence of sodium ethoxide in ethanol (**Figure 6**) [16].

2.1.7 Direct condensation of aldehydes and anthranilamide and its derivatives

Condensation of anthranilamide with aryl, alkyl and hetero-aryl aldehydes in refluxing ethanol in the presence of $CuCl_2$ generated the Schiff base intermediate **18**, which, in turn, was converted into the 2-substituted quinazolinones **4** in excellent yield (**Figure 7**). In a one-pot procedure, the aldehyde, anthranilamide and 3 equiv. of $CuCl_2$ were heated under reflux in ethanol for 2–3 h. After purification by chromatography, the 2-substituted quinazolinones **4** were isolated in 71–88% yield [17].

2.2 3-Substituted-4(3*H*)-quinazolines

2.2.1 Vilsmeier reagent in quinazolinone synthesis

3-Substituted-4(3*H*)-quinazolinone derivatives **22** have been prepared by treating 5-substituted-2-aminobenzoic acid derivatives **1** with the Vilsmeier

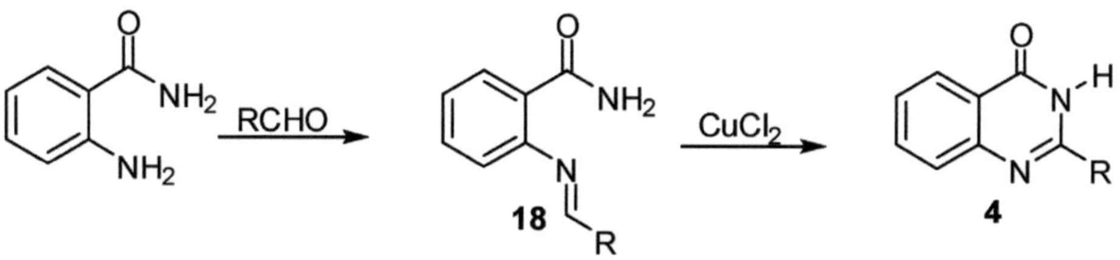

Figure 7.
Synthesis of 2-substituted-4(3H)-quinazolinones 4.

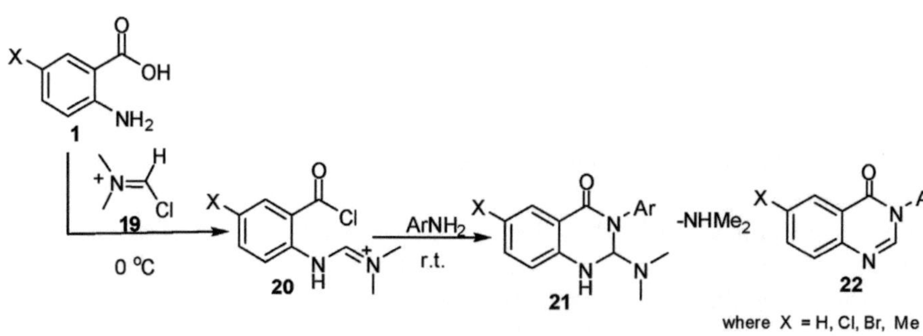

Figure 8.
Synthesis of 3-substituted-4(3H)-quinazolinones 22.

reagent **19** [18] to give the corresponding acid chlorides **20**. When the external amine was added to the reaction mixture at low temperature, the added amine reacted with the acid chloride **20** with subsequent loss of HCl followed by cyclisation to afford the intermediate **21**, which then expels a dimethyl amino group to afford the appropriately substituted quinazolinone **22** (**Figure 8**).

2.2.2 Via benzoxazinones

The (4*H*)-3,1-benzoxazin-4-one **23** was reacted with amines under reflux to afford the quinazolinones **24** good yields (**Figure 9**) [19].

2.2.3 Via metal salts as catalyst

4(3*H*)-Quinazolinones **22** have been synthesized in high to excellent yields through the one-pot condensation of anthranilic acid **1**, trimethyl orthoformate and primary amines in the presence of 5 mol % of bismuth (III) trifluoroacetate ($Bi(TFA)_3$) immobilized on *n*-butylpyridinium tetrachloroferrate ([nbp] $FeCl_4$) as ionic liquid (**Figure 10**) [20]. Also, the one-pot reaction was carried out in the presence of lanthanum (III) nitrate hexahydrate or *p*-toluenesulfonic acid [21].

2.3 2,3-Disubstituted-4(3*H*)-quinazolines

2.3.1 Formation of 2,3-disubstituted quinazolinones via benzoxazinone

Benzoxazinones are well-known as common intermediates in the synthesis of 2,3-disubstituted quinazolinones **24**. The most common approaches to synthesize

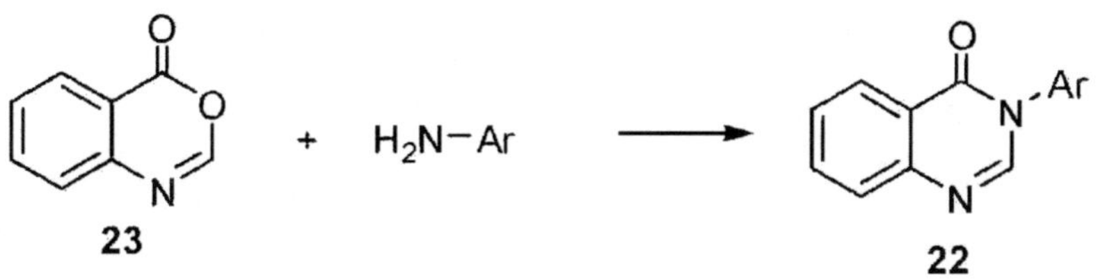

Figure 9.
Synthesis of 3-substituted-4(3H)-quinazolinones 22.

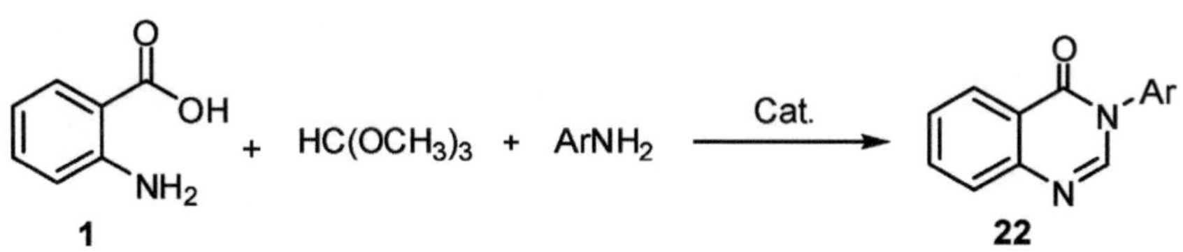

Figure 10.
Synthesis of 3-substituted-4(3H)-quinazolinones 22.

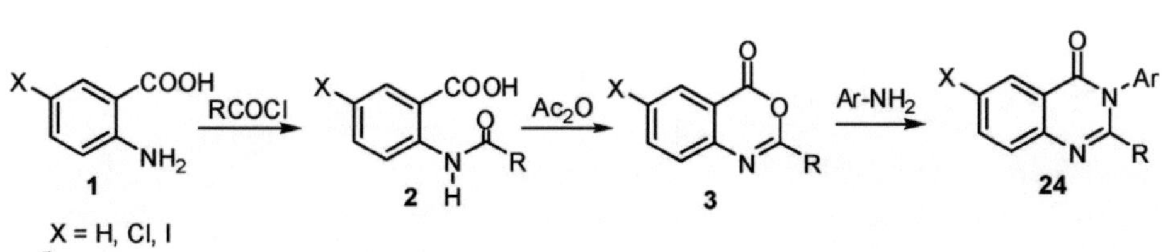

Figure 11.
Synthesis of 2,3-disubstituted quinazolinones 24.

3,2-disubstituted-4(3*H*)-quinazolinone derivatives involve amidation of anthralinic acid derivatives **1** and then treatment of the amidated anthranilic acid derivatives **2** with acetic anhydride to afford 3,1-benzoxazin-4-ones **3** in good yield, followed by condensation with nitrogen nucleophiles such as aromatic amines [22–24] or heterocyclic amines [25] (**Figure 11**).

Condensation of benzoxazinone **3** with hydrazine hydrate in *n*-butanol afforded 3-amino-quinazolinones **25** in good yield (**Figure 12**) [26, 27]. Also, bis-quinazolinones **26** were prepared by condensation of two moles of benzoxazinones **3** with one mole of a diamine under reflux [28].

2.3.2 Formation of 2,3-disubstituted quinazolinones via amide derivatives

Xue et al. reported the optimisation of Grimmel's conditions for generating 2,3-disubstituted-4(3*H*)-quinazolinones [29]. Thus, when amide derivatives **2** were heated with anilines in toluene or xylene in the presence of dehydrating agents such as phosphorous trichloride, phosphorous oxychloride or thionyl chloride, quinazolinone derivatives **24** were afforded. Benzenesulfonyl chloride was employed as coupling agent [30]. Treatment of amide derivatives **2** with hydrazine hydrate afforded 3-aminoquinazolinone derivatives **25** (**Figure 13**).

2.3.3 Combinatorial approach to quinazolinones

Traceless and chemoselective approach for the solid-phase synthesis of 2-arylamino-substituted quinazolinones with the possibility of manipulation at three positions has been developed by Yu et al. [31]. The nitro group at compounds **27** was subjected to reduction using tin(II) to afford the *ortho*-amino derivatives **28** and subsequently reacted with some aryl isothiocyanates to give thiourea derivatives **29**. The thiourea derivatives **29** were subjected to react at first with 2-chloro-1-methyl pyridinium iodide (Mukaiyama's reagent) and then reacted with the primary amines to afforded the guanidine derivatives **30**. When the guanidine derivatives **30** were subjected to intramolecular cyclisation using hydrofluoric acid, the desired 2-amino-substituted quinazolines **31** were obtained in good yields (**Figure 14**).

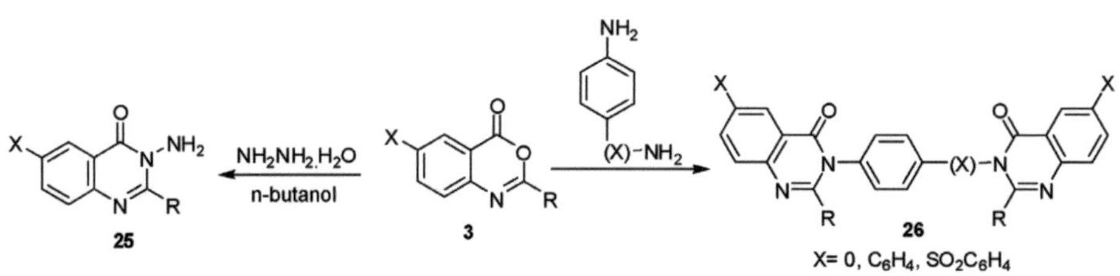

Figure 12.
Synthesis of 3-amino-quinazolinones 25 and bis- quinazolinones 26.

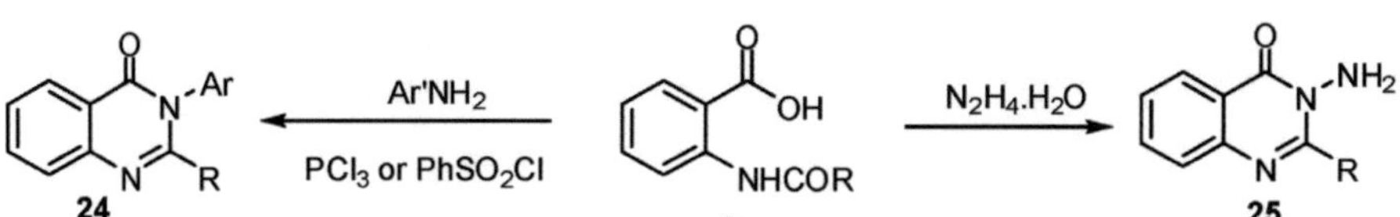

Figure 13.
Synthesis of 3-aminoquinazolinone derivatives 25.

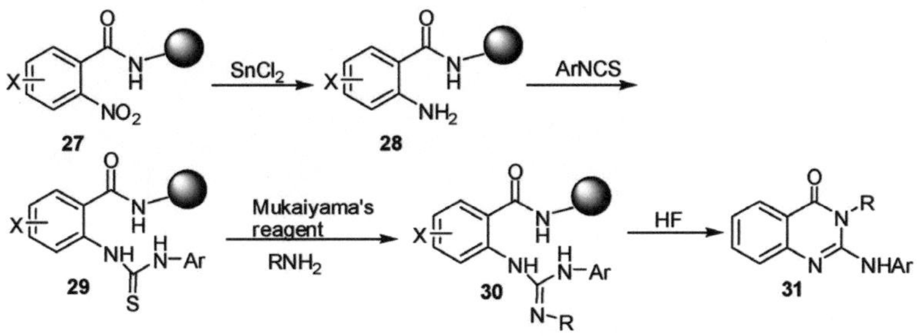

Figure 14.
Synthesis of 2-amino-substituted quinazolines 31.

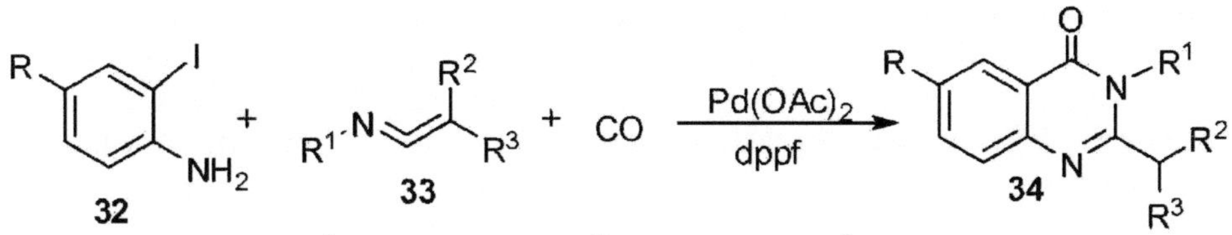

Figure 15.
Synthesis of 2-alkyl-4(3H)-quinazolinones 34.

2.3.4 Quinazolinone derivatives via palladium-catalyzed cyclocarbonylation

Cyclocarbonylation of *o*-iodoanilines **32** with ketenimine **33** using a palladium acetate/diphenylphosphinoferrocene catalyst was employed under a carbon monoxide pressure to afford the 2-alkyl-4(3*H*)-quinazolinones **34** in good to excellent yields (**Figure 15**) [32].

2.3.5 Chemoselective lithiation of quinazolinone derivatives

By direct lithiation of the 2-unsubstituted quinazolinone **22,** it was possible to carry out a range of electrophilic substitutions [19]. Also, chemoselective lithiation of 3-(acylamino)-quinazolines **35** was obtained by using of LDA where the reaction was regioselective at position 2. The similar phenomenon was observed with the corresponding 2-methyl quinazolines [33]. Reactions of the dilithio reagents with a range of electrophiles resulted in the production of the corresponding 2-substituted-4(3*H*)-quinazolinone derivatives **37** (**Figure 16**).

2.3.6 Formation of 2,3-disubstituted quinazolinones via isatoic anhydride

A more attractive and atom-efficient strategy for the preparation of 2,3-dihydroquinazolin-4(1*H*)-ones **38** was reported, which involved a one-pot three-component

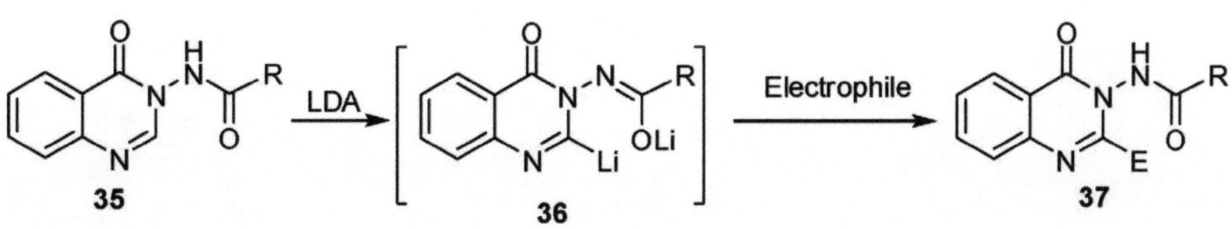

Figure 16.
Synthesis of 2-substituted-4(3H)-quinazolinone derivatives 37.

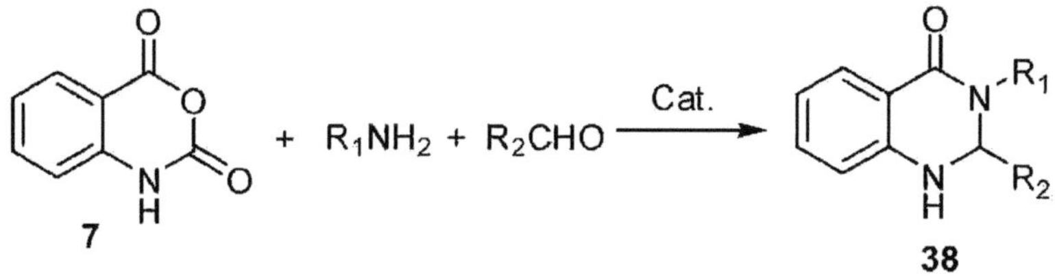

Figure 17.
*Synthesis of 2,3-dihydroquinazolin-4(1*H*)-ones **38**.*

reaction of isatoic anhydride 7, aldehydes and amines (**Figure 17**) [1, 34]. Multicomponent reactions or one-pot syntheses are attractive synthetic strategies, where the diversity may be achieved and the products are formed in a single step. A new method for the synthesis of 2-substituted-3-(phenylamino)-dihydroquinazolin-4(1*H*)-ones was developed; isatoic anhydride, phenylhydrazine and aldehyde using bentonite as catalyst in aqueous media under ultrasonic irradiation. This procedure showed good functional group tolerance [35].

3. Physical properties of 4(3H)-quinazolinones

3.1 Stability and tautomeric phase

4(3*H*)-Quinazolinones are stable to mild acid and alkaline treatment. They can be sublimated, and their parent substances can be redistillated. Weddige [36] recognized the tautomeric properties of 4(3*H*)-quinazolinones which could exit in three tautomeric forms **39a**, **39b** and **39c** (**Figure 18**). The presence of 4-hydroxy form **39b** was shown by its stability in aqueous alkali at pH 12 to give the anion. 4-Quinazolinones **39a** and **39c** are insoluble in alkali when a substitute is present on N^1 or N^3.

3.2 IR spectra

The IR spectra of 4(3*H*)-quinazolinone is characterized by a strong carbonyl band **39a** and **39c** at 1681 cm^{-1} and the N–H stretching band at 3402 cm^{-1} (inflection). Methyl groups in positions 2 and 3 have nearly the same effect in causing the carbonyl frequency to be lowered by 20–30 cm^{-1}. While methyl group at positions 1 and 2 lowered the frequency by 67 cm^{-1}, this large change was attributed to the presence of the β-double bond which is conjugated with the carbonyl group [37].

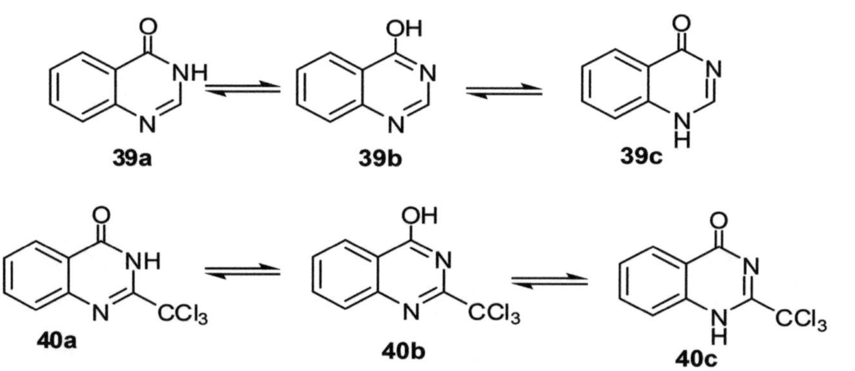

Figure 18.
*Tautomaric properties of 4(3*H*)-quinazolinones **39** and **40**.*

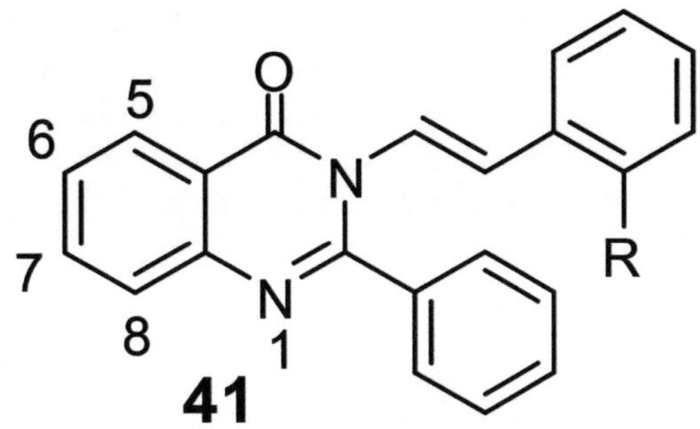

41

Figure 19.
Numbering of 4(3H)-quinazolinone 41.

3.3 UV spectra

The apparent dissociation constants of 4(3H)-quinazolinone **39** and 2-substituted-4(3H)-quinazolinones **40** were determined (**Figure 18**). Also, Ab initio quantum chemical calculations were performed for all possible tautomeric and protonation. The observed UV spectra revealed the number of dissociation constants. They concluded that the curves of 4(3H)-quinazolinones could be accounted for a mixture of tautomers due to the mobil hydrogen atom on N-3. The results showed a good correlation between experimentally determined pK_a values and theoretically calculated energies [38].

3.4 NMR spectroscopic studies of 4(3H)-quinazolinones

The NMR assignments of compound **41** (**Figure 19**) was based on simple ^1H and ^{13}C measurements and corroborated by ^1H-^1H COSY, gradient-enhanced ^{13}C-^1H HSQC and ^{13}C-^1H HMBC experiments. The ^1H NMR spectrum showed the aromatic protons in the range 7.27-8.56 ppm. From the 2D spectra, the signal assignments were thus at δ values: 8.56 (H5), 7.61 (H6), 7.65 (H7) and 7.27 ppm (H8) [39].

4. Chemical reaction of 4(3*H*)-quinazolinones

4.1 Reactivity of the 2-methyl group

4.1.1 Oxidation

The methyl group in the structure of 2-methyl-4(3*H*)-quinazolinones **24** has possibility for oxidation to generate further useful fictionalization at this position [22–24]. So, by using SeO_2 as oxidant agent, the methyl group in **24** was converted to

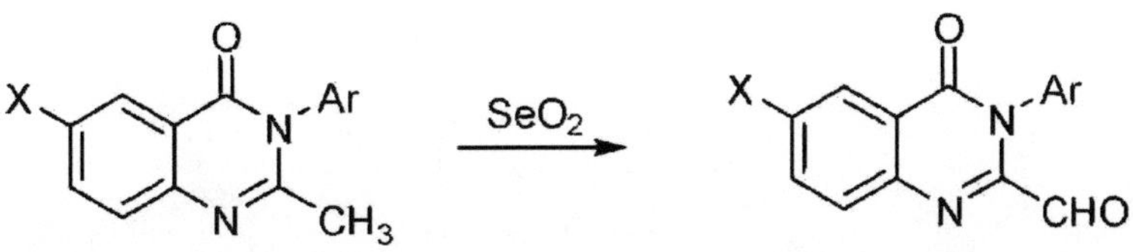

24
42

Figure 20.
Synthesis of 4(3H)-quinazolinone-2-carboxaldehydes 42.

formyl group, and the novel 4(3*H*)-quinazolinone-2-carboxaldehydes **42** were furnished and subjected to further reaction to give new quinazoline derivatives having a azomethine, oxazolone, imidazolidine, pyrazolidine, pyridine, pyrimidine and variously substituted C-2. Also, series of 3-aryl-4(3*H*)-quinazolinone-2-carboxaldehyde thiosemicarbazones were synthesized via condensation of the 4(3*H*)-quinazolinone-2-carboxaldehydes **42** with the desired thiosemicarbazide derivatives (**Figure 20**).

4.1.2 Reaction with aldehydes

The 2-methyl group in substituted 4(3*H*)-quinazolinone is reactive as shown by the ease of its condensation with aldehydes to give the corresponding 2-styryl derivatives. 6-Chloro-2-methyl-quinazolin-4(3*H*)-one was refluxed for 12 h in glacial acetic acid with pyridine-2-carbaldehyde to give 6-chloro-2-(2-pyridin-2-yl-vinyl)-4(3*H*)-quinazolinone [7]. Also, a series of 3-[5-substituted phenyl-1, 3, 4-thiadiazole-2-yl]-2-styryl-4(3*H*)-quinazolinones **43** were synthesized by refluxing equimolar amount of 3-(1′3′4′-thiadiazolyl)-2-methyl quinazoline and aromatic aldehyde in glacial acetic acid [40] (**Figure 21**).

4.1.3 Bromination

The methyl group in 2-methyl-3-aryl-quinazoline has been found to undergo bromination by bromine to give bromomethyl compound **44** (**Figure 22**). Many compounds were synthesized via treatment compound **44** with potassium salts of organic compounds [40]. Other compounds were synthesized via treatment compound **47** with amine [41].

4.1.4 Lithiation

2-Methyl-4(3*H*)-quinazolinone **45** underwent fold metalation with alkyl lithium to form lithio salt **46** which react with electrophilies (methyl iodide, ethyl iodide, allyl bromide, benzyl chloride, etc.) exclusively at the exocyclic carbanion site to produce quinazolinone derivatives **47** (**Figure 23**) [42–44].

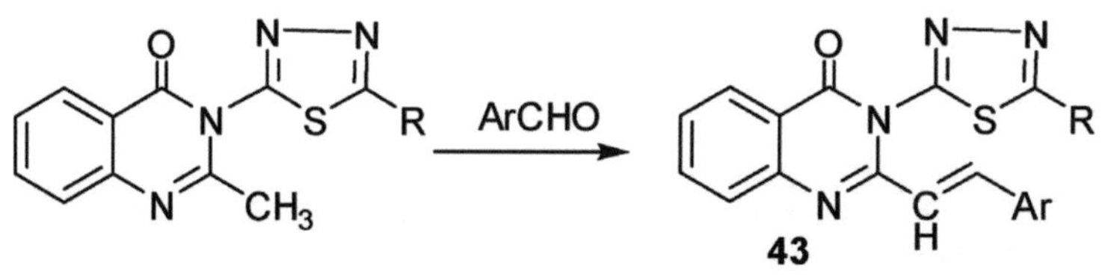

Figure 21.
*Synthesis of 2-styryl-4(3*H*)-quinazolinones **43**.*

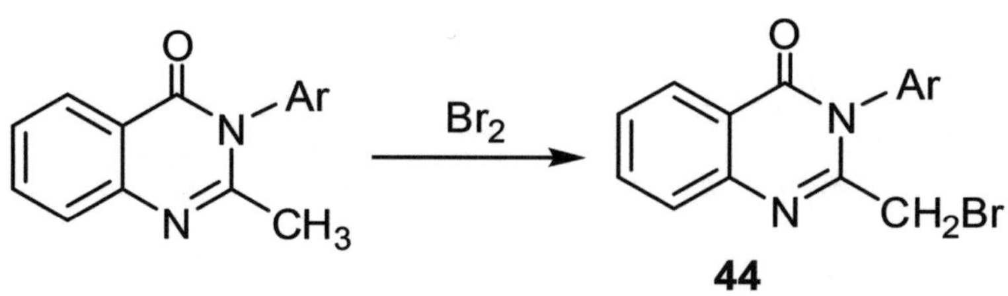

Figure 22.
*Synthesis 2-bromomethyl-3-aryl-quinazolines **44**.*

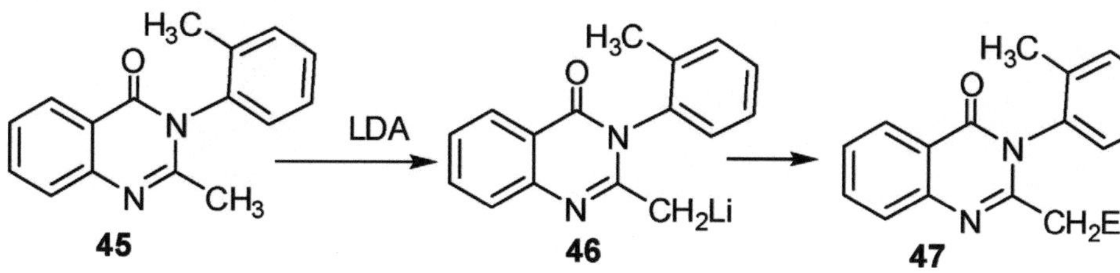

Figure 23.
Synthesis 2-substitutedmethyl-3-aryl-quinazolines 47.

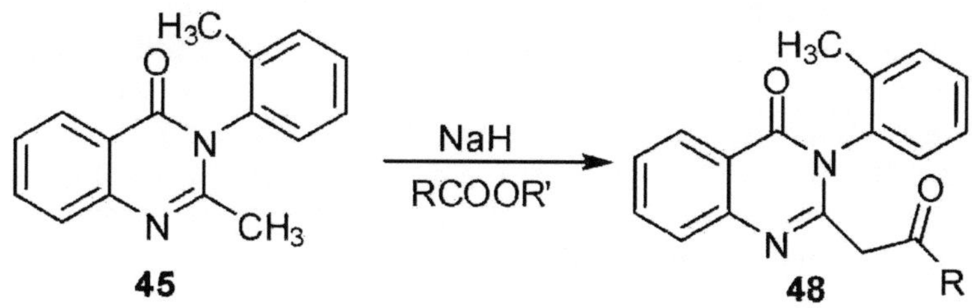

Figure 24.
Synthesis 2-substitutedmethyl-3-aryl-quinazolines 48.

4.1.5 Acylation

A series of 4(3*H*)-quinazolinones **48** [45] structurally related to methaqualone (2-methyl-3-o-tolyl-4(3*H*)-quinazolinone) were synthesized and evaluated for anticonvulsant activity. They prepared by treating 2-methyl-3-aryl-4(3*H*))-quinazolinone **45** with sodium hydride followed by the appropriate methyl or ethyl ester (**Figure 24**).

4.2 Reactivity of the 3-amino group

4.2.1 Synthesis of cyanoacetamide derivative synthons

The thermal fusion of 3-amino-4(3*H*)-quinazolinone **49** with ethyl cyano-acetate afforded cyanoacetamide derivatives **50** (**Figure 25**). Cyanoacetamide derivatives **50** are highly reactive, polyfunctional compounds that possess both electrophilic and nucleophilic centres. Cyanoacetamide derivative **50** was widely used as an active synthon for the syntheses of many open-chain systems and poly-substituted heterocyclic compounds. The chemical properties of cyanoacetamide derivative **50** have been used to design various heterocyclic moieties with different ring sizes [26].

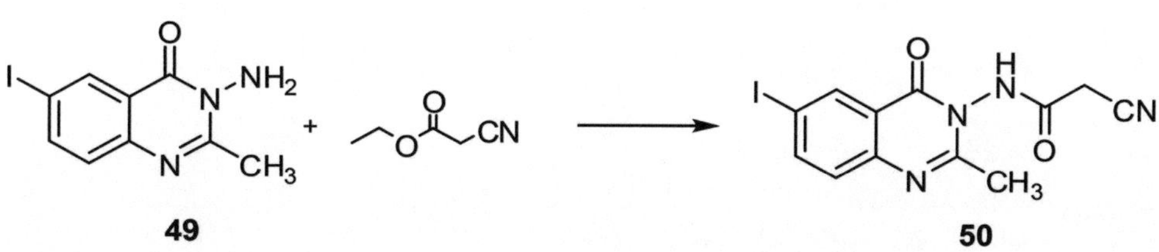

Figure 25.
Synthesis of cyanoacetamide derivatives 50.

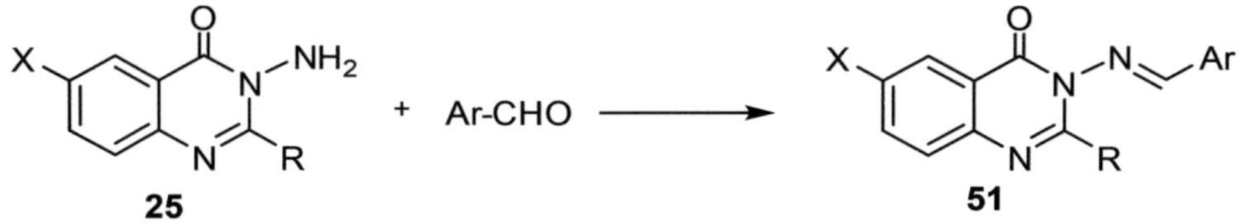

Figure 26.
Synthesis of Schiff's bases 51.

4.2.2 Condensation with aldehydes

A series of Schiff's bases **51** were prepared essentially by the usual condensation reaction between the 3-amino-quinazolinone derivative **25** and the aldehydes (**Figure 26**). On the other hand, when two moles of compound **25** were treated with one mole of terephthaldehyde in ethanol under reflux, the polycyclic compound was obtained as bis-quinazolinones. Another bis-quinazolinone was obtained when two moles of compound **25** was treated with one mole of ethyl *tere*-phthalate in dimethylformamide under reflux conditions [27, 46].

4.2.3 Acylation and/or alkylation

Acylation and/or alkylation of 3-amino-4(3*H*)-quinazolinones **25** using ethyl chloroformate, ethyl chloroacetate, chloro acetylchloride and ethyl acetoacetate in proper solvent afforded 3-(*N*-acyl/aroylamino)-2-methyl-4(3*H*)-quinazolinone derivatives **52** [47] (**Figure 27**).

4.2.4 Reaction with isocyanate and isothiocyanates

Some new urea and thiourea derivatives **53** were synthesized by treatment of 3-amino-4(3*H*)-quinazolinones **25** with isocyanates and isothiocyanates [28] (**Figure 28**).

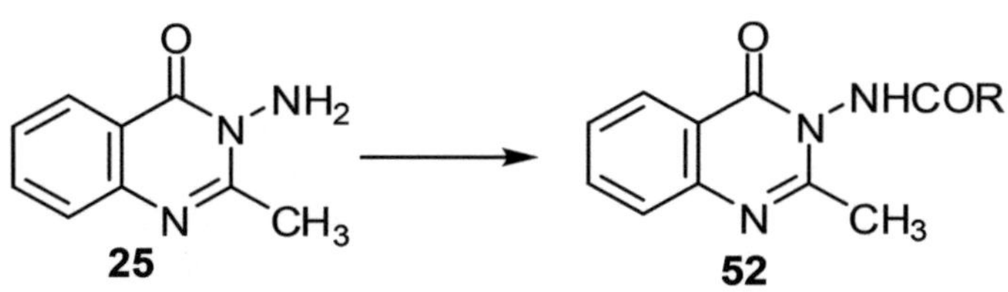

Figure 27.
*Acylation or alkylation of 3-amino-4(3*H*)-quinazolinones 25.*

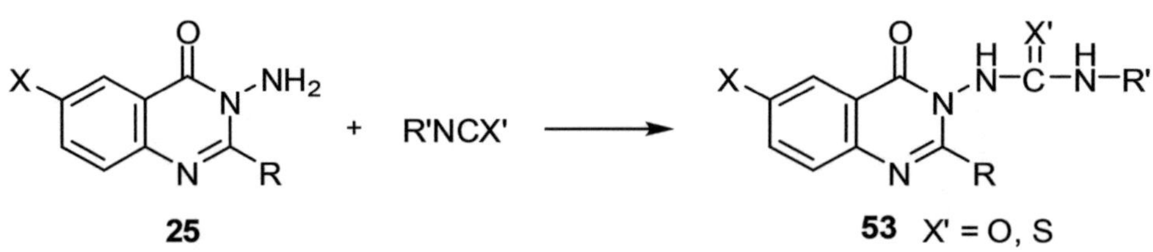

Figure 28.
Synthesis of urea and thiourea derivatives 53.

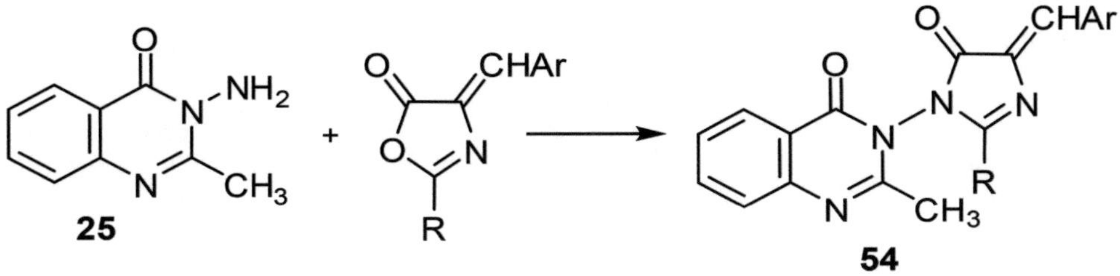

Figure 29.
Synthesis of imidazole derivatives 54.

4.2.5 Reaction with oxazole

Condensation of 3-aminoquinazoline **25** with oxazole derivatives afforded imidazole derivatives **54** [47] (**Figure 29**).

4.3 Electrophilic substitution

4.3.1 Nitration

Nitration of 4(3*H*)-quinazolinone **39** with fuming nitric acid and sulphuric acid afforded 6-nitro-4(3*H*)-quinazolinone derivative **55** [48] (**Figure 30**).

4.3.2 Chlorination

Heating of 4(3H)-quinazolinones **56** with chlorination agent afforded 4-chloroquinazolines **57** [49]. Chlorination agent was phosphoryl chloride alone or a mixture of phosphorus pentachloride and phosphoryl chloride, other chlorinating agents such as thionyl chloride or phosgene was used for chlorination of quinazolinones (**Figure 31**).

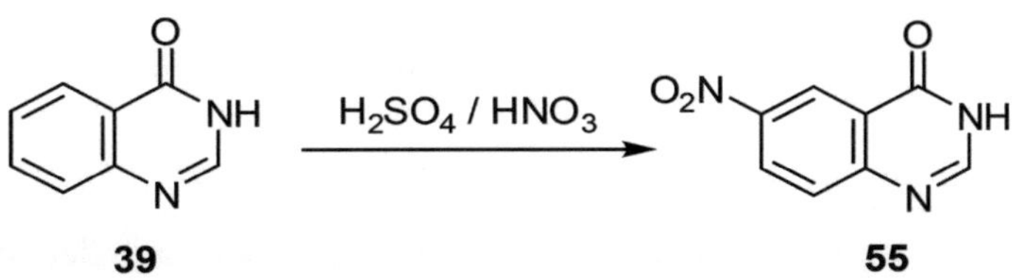

Figure 30.
Nitration of 4(3H)-quinazolinone 39.

Figure 31.
Chlorination of 4(3H)-quinazolinone derivatives 56.

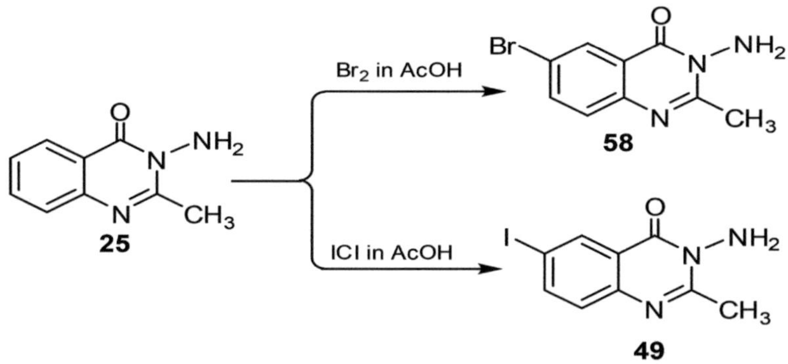

Figure 32.
Synthesis of 6-bromoquinazolinone 58 and 6-iodoquinazolinone 49.

4.3.3 Bromination

The use of bromine in acetic acid for direct bromination of 3-amino-2-methylquinazolin-4(3*H*)-one **25** has been reported for the formation of 6-bromo-3-amino-2-methylquinazolin-4(3*H*)-ones **58** (**Figure 32**) [50].

4.3.4 Iodination

Treatment of 3-amino-2-methylquinazolin-4(3*H*)-one **25** with iodine monochloride in acetic acid afforded the corresponding 6-iodo-3-amino-2-methylquinazolin- 4(3*H*)-one **49** in high yields [50] (**Figure 32**).

4.4 Reaction of 4(3*H*)-quinazolinones with metal ions

3-Amino-4(3*H*)-quinazolinones **25** possess coordinating sites and they were applied to form complexes **59** and bis-complexes **60** with different metal ions (**Figure 33**) [51–53]. Also, considerable attention has been directed to the chemistry of their Schiff's bases **51** and **61** [27], where the Schiff's base complexes of 4(3*H*)-quinazolinones **62-65** were prepared and characterized. The complexes of metal ions with 2-substituted-3-anilino-4(3*H*) quinazolinone were prepared [54]. Moreover, the complexes of Cu (II), Co (II), Zn (II) and Cd (II) with 2-methyl-3-hydroxy-4(3*H*)-quinazolinone and 2-methyl-3-pyridinyl-4(3*H*)-quinazolinone have been prepared [55]. The analytical and spectral data indicate these ligands act as bidentate and the metal complexes are octahedral, tetragonal, square planer and tetrahedral [56].

Thiosemicarbazones can be reacting with metallic cations to give metal complexes. Thiosemicarbazones **66** as the ligands act as chelating agents which were bonding through the sulfur and azomethene nitrogen atoms (**Figure 34**). So, when metal salts such as $CuCl_2$ or $ZnCl_2$ were treated with the thiosemicarbazone derivatives **66** (0.01 mole) in dioxane the corresponding complexes **67** were obtained in good yield. On the other hand, the corresponding biscomplexes **68** were afforded when a solution of $CuCl_2$ or $ZnCl_2$ (0.01 mole) was added to a stirred solution of thiosemicarbazone derivatives (0.02 mole) in dioxane at reflux temperature [22].

4.5 Cycloaddition reaction

Reaction of quinazolinone derivatives **69** with malononitrile gave pyrrolo-quinazolinones **70** [57] (**Figure 35**). Several new pyrrolo-quinazolinone derivatives were synthesized via a novel rote involving the action of dipolarophiles on the diionic species generated in situ from the reaction of N-chlorosuccinimide with 2-methylquinazolin-4-one and subsequent treatment with triethyl amine [58].

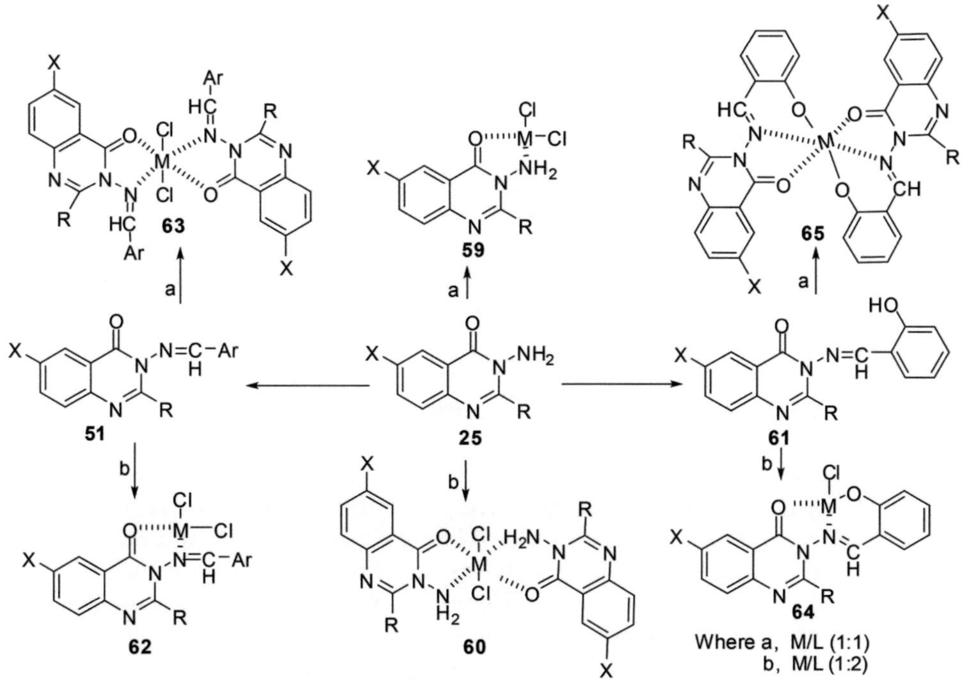

Figure 33.
*Reaction of 4(3H)-quinazolinones **25**, **51** and **61** with metal ions.*

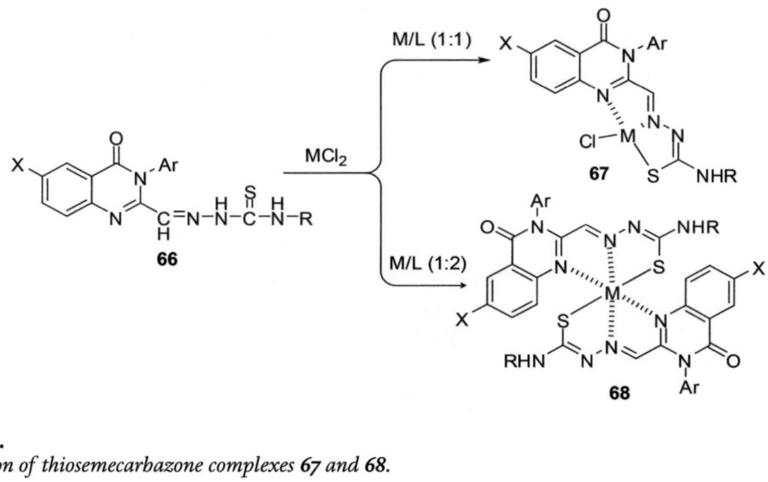

Figure 34.
*Preparation of thiosemecarbazone complexes **67** and **68**.*

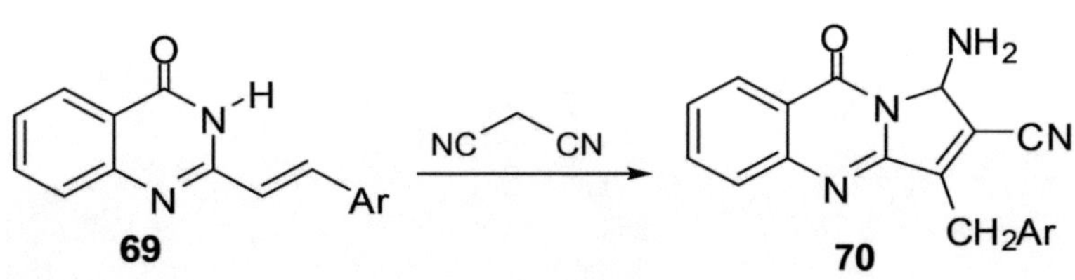

Figure 35.
*Synthesis of pyrrolo-quinazolinones **70**.*

4.6 Action of phosphorous sulphide

Treating of 4(3*H*)-quinazolinone derivatives **24** with P_2S_5 or phosphorus decasulfide in pyridine afforded the corresponding 2-methyl-3-aryl-quinazoline-4(3*H*)-thiones **71** [59] (**Figure 36**).

5. Biological properties

Quinazoline is the building stone for many naturally occurring alkaloids [60]. Many 4(3*H*)-quinazolinone derivatives represent an important category among heterocyclic compounds of medicinal interest. Other derivatives of 4(3*H*)-quinazolinones possess a wide range of biological activities especially on the central nervous system. Moreover, other quinazoline derivatives have been reported for their broad-spectrum biological activities as herein illustrated.

5.1 4(3*H*)-Quinazolinone derivatives as antitumour

Structure modification of folic acid led to the discovery of a number of antifolates as efficient anticancer agents. For example, Raltitrexed has been registered for the treatment of cancer [61]. Many quinazolinone derivatives with side chains have been reported to exhibit significant inhibitory activity against tumor cells [62]. The 2-substituted mercapto-4(3*H*)-quinazolinone bearing 6-iodo and 2-heteroarylthio is identified as active anticancer agent [63].

5.2 4(3*H*)-Quinazolinone derivatives as sedative hypnotic agents

The designation of the sedative hypnotic activity of 4(3*H*)-quinazolinones led to the discovery of methaqualone as nonbarbiturate hypnotic agent. In 1965, methaqualone was introduced as sleeping pills (nonaddictive, nonbarbiturate) under the trade name Quaalude. Due to the abuse of methaqualone, it is banned in most countries [64].

5.3 4(3*H*)-Quinazolinone derivatives as anticonvulsant agents

The search for new antiepileptic drugs with reduced toxicity and lower side-effects is continuous. 4(3*H*)-Quinazolinone represents a very good nucleus for preparation of some new sedative/hypnotic and anticonvulsant agents, since such a heterocyclic system possesses the pharmacophoric moiety. From the literature survey, it was found that the 3*H*-quinazolin-4-one has been reported to possess different pharmacological effects, namely, sedative-hypnotic and anticonvulsant ones [65].

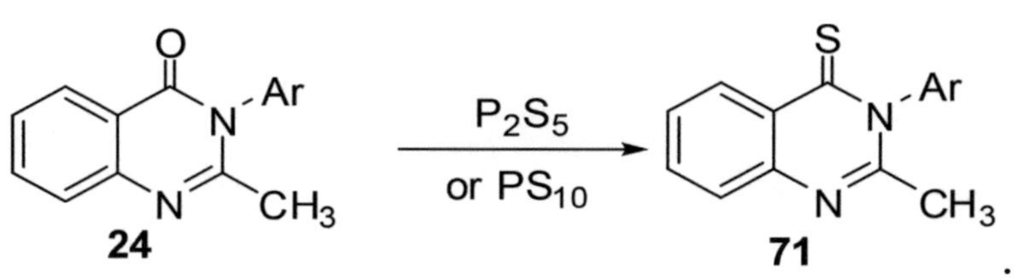

Figure 36.
*Synthesis of 2-methyl-3-aryl-quinazoline-4(3H)-thiones **71**.*

5.4 4(3*H*)-Quinazolinone derivatives as antimicrobial agents

A large number of quinazolinone derivatives have been synthesized and screened for their antimicrobial activities, and some of them showed their efficacy [66]. Also, quinazolinones metal complexes were synthesized, and their antimicrobial activities were screened. It is observed that the ligands exhibited less fungicidal activities than their complexes. Also the antibacterial activities were increased when quinazolines were complexed with metals.

5.5 4(3H)-Quinazolinone derivatives as anti-inflammatory agents

Thiadiazolyl-3-amino-4(3*H*)-quinazolinone derivatives was prepared in quantities yields. The products were evaluated for anti-inflammatory properties by std. in vivo and in vitro models, and they exhibited significant protection against carrageenan-induced rat paw oedema [67].

5.6 4(3H)-Quinazolinone derivatives as diuretic agents

Some 4(3*H*)-quinazolinone derivatives bearing thiazole or 1,3,4-thiadiazole moieties were prepared due to their expected diuretic activity. Some of them showed significant diuretic activity [68].

6. Conclusion

A large number of compounds which contain quinazoline moiety are known in medicinal chemistry as important compounds for their therapeutic value. Recently, there has been an increased interest in the chemistry of 4(3H)-quinazolinone system. Many derivatives of this system showed analgesic, anti-inflammatory, antiulcer, anticonvulsant, antibacterial, antifungal, anticancer and antiproliferative activities. The most common approaches to synthesize 3,2-disubstituted-4(3H)-quinazolinone derivatives involve the following steps: the amidation of 2-aminobenzoic acid derivatives and treatment of amidated anthranilic acid derivatives with acetic anhydride (or acid chloride) to afford the benzoxazinone, followed by their condensation with nitrogen nucleophiles.

Author details

Samir Y. Abbas
Organometallic and Organometalloid Chemistry Department, National Research Centre, Cairo, Egypt

*Address all correspondence to: samiryoussef98@yahoo.com

References

[1] Abbas SY, El-Bayouki KAM, Basyouni WM. Utilization of isatoic anhydride in the syntheses of various types of quinazoline and quinazolinone derivatives. Synthetic Communications. 2016;**46**(12):993-1035

[2] Safari J, Gandomi-Ravandi S. Silver decorated multi-walled carbon nanotubes as a heterogeneous catalyst in the sonication of 2-aryl-2,3-dihydroquinazolin-4(1*H*)-ones. RSC Advances. 2014;**4**:11654-11660

[3] Kim NY, Cheon C-H. Synthesis of quinazolinones from anthranilamides and aldehydes via metal-free aerobic oxidation in DMSO. Tetrahedron Letters. 2014;**55**:2340-2344

[4] Cheng R, Tang L, Guo T, Zhang-Negrerie D, Du Y, Zhao K. Oxidant- and metal-free synthesis of 4(3*H*)-quinazolinones from 2-amino-*N*-methoxybenzamides and aldehydes *via* acid-promoted cyclocondensation and elimination. RSC Advances. 2014;**4**:26434-26438

[5] Rao R, Raghunadh A, Mekala R, Meruva S, Pratap T, Krishna T, et al. Glyoxylic acid in the reaction of isatoic anhydride with amines: A rapid synthesis of 3-(un)substituted quinazolin-4(3H)-ones leading to rutaecarpine and evodiamine. Tetrahedron Letters. 2014;**55**:6004-6006

[6] Santagati N, Bousquet E, Spadaro A, Ronsisvalle G. 4-Quinazolinones:Synthesisandreduction of prostaglandin E2 production. Il Farmaco. 1999;**54**:780-784

[7] Jiang JB, Hesson DP, Dusak BA, Dexter DL, Kang GL, Hamel E. Synthesis and biological evaluation of 2-styrylquinazolin-4(3H)-ones, a new class of antimitotic anticancer agents which inhibit tubulin polymerization. Journal of Medicinal Chemistry. 1990;**33**:1721-1728

[8] Rad-Moghadam K, Mohseni M. An expeditious and solvent-free route to the synthesis of 2-substituted quinazolin-4(3H)-ones under microwave conditions. Journal of Chemical Research. 2003;**35**:487

[9] Li F, Feng Y, Meng Q, Li W, Li Z, Tao F. An efficient construction of quinazolin-4(3H)-ones under microwave irradiation. Arkivoc. 2007;**i**:40-50

[10] Baker BR, Almaula PI. Nonclassical antimetabolites. X. A facile synthesis of 4-quinazolone-2-carboxylic acid and the structure of bogert's ammonium salt. The Journal of Organic Chemistry. 1962;**27**:4672-4674

[11] Hennequin LF, Boyle FT, Wardleworth JM, Marsham PR, Kimbell R, Juckman AL. Quinazoline antifolates thymidylate synthase inhibitors: Lipophilic analogues with modification to the C2-methyl substituent. Journal of Medicinal Chemistry. 1996;**39**:695-704

[12] Connolly DJ, Guiry PJ. A facile and versatile route to 2-substituted-4(3*H*)-quinazolinones and quinazolines. Synlett. 2001;**11**:1707-1710

[13] Yang RY, Kaplau A. A concise and efficient solid-phase synthesis of 2-amino-4(3H)-quinazolinones. Tetrahedron Letters. 2000;**41**:7005-7008

[14] Croce PD, Ferracioli R, La Rosa C. Reactivity of 1-Aryl-4-dimethylamino-2-phenyl-1,3-diaza-1,3-butadienes towards Dienophiles, 1,3-Dipoles and Carbanions. Heterocycles. 1997;**45**:1309-1318

[15] Couture A, Cornet H, GrandclaudonP.Anexpeditioussynthesis of 2-aryl- and 2-alkylquinazolin-4(3*H*)-ones. Synthesis. 1991;**1991**:1009-1010

[16] Hess HJ, Cronin TH, Scriabine A. Antihypertensive 2-amino-4(3H)-

quinazolinones. Journal of Medicinal Chemistry. 1968;**11**:130-136

[17] Abdel-Jalil RJ, Volter W, Saeed M. A novel method for the synthesis of 4(3H)-quinazolinones. Tetrahedron Letters. 2004;**45**:3475-3476

[18] Majo VJ, Perumal PT. Dimerization of substituted 2-aminobenzoic acids under Vilsmeier conditions: A novel route to the synthesis of 4-(3H)-quinazolinones. Tetrahedron Letters. 1996;**37**:5015-5018

[19] Dai X, Virgil S. Synthesis of 2-heterosubstituted quinazolinone atropisomeric phosphine ligands by direct lithiation of a 2-unsubstituted quinazolinone system. Tetrahedron: Asymmetry. 1999;**10**:25-29

[20] Khosropour AR, Baltork IM, Ghorbankhani H. Bi(TFA)3-[nbp] $FeCl_4$: A new, efficient and reusable promoter system for the synthesis of 4(3H)-quinazolinone derivatives. Tetrahedron Letters. 2006;**47**:3561-3564

[21] Narasimhulu MJ, Mahesh KC, Reddy TS, Venkalesewarlu Y. Lanthanum(III) nitrate hexahydrate or p-toluenesulfonic acid catalyzed one-pot synthesis of 4(3H)-quinazolinones under solvent-free conditions. Tetrahedron Letters. 2006;**47**:4381-4383

[22] El-Bayouki KAM, Aly MM, Mohamed YA, Basyouni WM, Abbas SY. Synthesis and biological activity of some 4(3H)-quinazolinone-2-carboxaldehyde thiosemicarbazone and their metal complexes. European Journal of Medicinal Chemistry. 2010;**45**:3365-3373

[23] Farag AA, Khalifa EM, Sadik NA, Abbas SY, Al-Sehemi AG, Ammar YA. Synthesis, characterization and evaluation of some novel 4(3H)-quinazolinone derivatives as anti-inflammatory and analgesic agents. Medicinal Chemistry Research. 2013;**22**:440-452

[24] Abbas SY, El-Bayouki KAM, Basyouni WM, Mostafa EA. New series of 4(3H)-quinazolinone derivatives: Syntheses and evaluation of antitumor and antiviral activities. Medicinal Chemistry Research. 2018;**27**:571-582

[25] Párkányi CJ, Schmidt DS. Synthesis of 5-chloro-2-methyl-3-(5-methylthiazol-2-yl)-4(3H)-quinazolinone and related compounds with potential biological activity. Journal of Heterocyclic Chemistry. 2000;**37**:725-729

[26] El-Bayouki KAM, Aly MM, Mohamed YA, Basyouni WM, Abbas SY. Novel 4(3H)-quinazolinone containing biologically active thiazole, pyridinone and chromene of expected antitumor and antifungal activity. European Journal of Chemistry. 2011;**2**(4):455-462

[27] El-Bayouki KAM, Aly MM, Mohamed YA, Basyouni WM, Abbas SY. Synthesis and biological activities of some 3-(amino or arylidine)-2-pyridinyl-4(3H)-quinazolinone and some of their complexes. Al-Azhar Bulletin of Science. 2010;**21**:1-20

[28] Ammar YA, Mohamed YA, El-Gaby MSA, Abbas SY. Synthesis of some biologically active 4(3H)-quinazolinones derived from 2,3-pyridine dicarboxylic anhydride. Chemical Sciences Journal. 2011;**2011**:1-10

[29] Xue S, McKenna J, Shieh W-C, Repic O. A Facile synthesis of C_2,N_3-disubstituted-4-quinazolone. The Journal of Organic Chemistry. 2004;**69**:6474-6477

[30] Dai X, Wong A, Virgil S. Synthesis and resolution of quinazolinone atropisomeric phosphine ligands. The Journal of Organic Chemistry. 1998;**63**:2597-2600

[31] Yu Y, Ostresh JM, Houghten RA. A traceless approach for the parallel

solid-phase synthesis of 2-(arylamino) quinazolinones. The Journal of Organic Chemistry. 2002;**67**:5831-5834

[32] Larksarp C, Alper H. Palladium-catalyzed cyclocarbonylation of *o*-iodoanilines with heterocumulenes: Regioselective preparation of 4(3*H*)-quinazolinone derivatives. The Journal of Organic Chemistry. 2000;**65**:2773-2777

[33] Smith K, El-Hiti G, Abbdel-Megeed M, Abdo A. Lithiation of 3-(acylamino)-2-unsubstituted-, 3-(acylamino)-2-ethyl-, and 3-(acylamino)-2-propyl-4(3H)-quinazolinones: Convenient syntheses of more complex quinazolinones. The Journal of Organic Chemistry. 1996;**61**:647-655

[34] Khatab TK, El-Bayouki KAM, Basyouni WM, El-Basyoni FA, Ali MM, Abbas SY, et al. A novel series of 2,3-dihydroquinazolin-4(1H)-one derivatives: Efficient one-pot synthesis and evaluation of antitumor and antiviral activity. Research Journal of Pharmaceutical, Biological and Chemical Sciences. 2015;**6**(1):281-291

[35] Ma Y, Ren D, Zhang J, Liu J, Zhao J, Wang L, et al. Synthesis, antibacterial activities evaluation, and docking studies of some 2-substituted-3-(phenylamino)-dihydroquinazolin-4(1H)-ones. Tetrahedron Letters. 2015;**56**:4076-4079

[36] Weddig A. Beiträge zur kenntniss des o-amidobenzamids, mitgetheilt von a. weddige. Journal für Praktische Chemie. 1887;**36**:141-154

[37] Culbertson H, Decius JC, Christensen BE. Quinazolines. XIII. A study of the infrared spectra of certain quinazoline derivatives. Journal of the American Chemical Society. 1952;**47**:4834-4838

[38] Sim Y-L, Omer N, Khan MN, Pratt LM. Ultravioletevisible study on acidebase equilibria of 2-substituted-4(3H)-quinazolinones. Tetrahedron. 2013;**69**:2524-2533

[39] Brahman D, Sinha B. Synthesis, characterization and antibacterial activities of Zn(II) and Cd(II) complexes of a 3-amino-2-phenylquinazolin-4(3H)-one Schiff base. Journal of the Serbian Chemical Society. 2014;**79**(12):1505-1513

[40] Jatav V, Mishra P, Kashaw S. Synthesis and CNS depressant activity of some novel 3-[5-substituted 1,3,4-thiadiazole-2-yl]-2-styryl quinazoline-4(3H)-ones. European Journal of Medicinal Chemistry. 2008;**43**(9):1945

[41] Hassan GS, Georgey HH, Ibrahim NA, Zaki SS. Synthesis of novel substituted quinazoline and quinazolin-4(3H)-one derivatives of expected antitumor activity. Der Pharma Chemica. 2015;**7**(7):106-119

[42] Sugimoto O, Yamauchi Y, Tanji K-I. Preparation of a useful synthetic precursor, 2-substituted 4(3H)-quinazolinone: Directed lithiation and N3-deprotection of 3-t-butoxycarbonyl-4(3H)-quinazolinone. Heterocycles. 2002;**57**:323-326

[43] Smith K, Hiti G, Abdel-Mageg MF, Abdo MA. Lithiation of 2-alkyl-3-amino- and 2-alkyl-3-(methylamino)-4(3H)-quinazolinones. The Journal of Organic Chemistry. 1996;**61**:656-661

[44] Rathman L, Sleevi CM, Krafft ME, Wolfe JF. Functionalization of 2-methyl-3-o-tolyl-4(3H)-quinazolinone and related compounds through carbanion reactions at the 2-methyl group. The Journal of Organic Chemistry. 1980;**45**:2169-2176

[45] Wolfe JF, Rathman TL, Sleevi MC, Campbell JA, Greenwood TD. Synthesis and anticonvulsant activity of some

new 2-substituted 3-aryl-4(3H)-quinazolinones. Journal of Medicinal Chemistry. 1990;**33**:161-166

[46] Madkour HMF. Reactivity of 4H-3,1-benzoxazin-4-ones towards nitrogen and carbon nucleophilic reagents: Applications to the synthesis of new heterocycles. Arkivoc. 2004;**I**:36-54

[47] Ibrahim SS, Abdel-Halim AM, GaberY, El-EdfawyS, Abdel-RahmanRM. Synthesis and Biological evaluation of some new fused quinazoline derivatives. The Journal of Chemical Research. 1997;**1041**:154

[48] Kabri Y, Verhaeghe P, Gellis A, Vanelle P. Regioselective Suzuki-Miyaura Reaction: Application to the Microwave-promoted Synthesis of 4,7-Diarylquinazolines. Molecules. 2010;**15**(5):2949-2961

[49] Takase Y, Saeki T, Watanabe N, Adachi H, Souda S, Saito I, et al. Phosphodiesterase inhibitors. 2. Requirement of 6-substitution of quinazoline derivatives for potent and selective inhibitory activity. Journal of Medicinal Chemistry. 1994;**37**(13):2106-2111

[50] Mudassar A, Yeshwant B. Synthesis of 6-iodo/bromo- 3-amino-2-methylquinazolin-4(3H)-ones by direct halogenation and their Schiff base derivatives. Arkivoc. 2006;**xi**:221-226

[51] Reddy KL, Chandraiah AR, Reddy KA, Lingaiah P. Spectral studies of platinum(II) complexes with 2,3-disubstituted quinazolin-(3H)-4-ones. Indian Journal of Chemistry -Section A. 1989;**28A**:622-626

[52] Prasad KS, Kumar LS, Chandan S, Jayalakshmi B, Revanasiddapp HD. Diorganotin(IV) complexes of biologically potent 4(3H)-quinazolinone derived Schiff bases: Synthesis, spectroscopic characterization, DNA interaction studies and antimicrobial activity. Spectrochimica Acta Part A: Molecular and Biomolecular Spectroscopy. 2011;**81**:276-282

[53] Reddy KL, Reddy KV, Lingaiah P. Ruthenium(II) complexes with 2-methyl-3-substituted (3H)-quinazolin-4-ones. Polyhydron. 1986;**5**:1519

[54] Reddy KL, Srihari S, Lingaiah P. Synthesis & characterization of complexes of copper(II), nickel(II), cobalt(II), zinc(II) & cadmium(II) with 2-methyl/phenyl-3-anilinoquinazoline-4-ones. Indian Journal of Chemistry. 1984;**23A**:172

[55] Reddy KL, Srihari S, Lingaiah P. Synthesis & characterization of complexes of Co(II), Ni(II), Cu(II), Zn(II) & Cd(II) with 2,3-disubstituted quinazolin-(3H)-4-ones. Indian Journal of Chemistry. 1985;**24A**:318-322

[56] Prabhakar B, Reddy K, Lingaiah P. Synthesis & spectral studies of complexes of Co(II), Ni(II), Cu(II), Zn(II), Ru(II), Pd(II) & Pt(II) with 2,3-disubstituted quinazolin-(3H)-4-ones. Indian Journal of Chemistry -Section A. 1988;**27A**:217-221

[57] Hishmat OH, Magd El-Din AA, Ismail NA. Synthesis of new substituted pyridines, pyrano[2,3-d]imidazoles and pyrrolo[2,1-b]quinazolines. Organic Preparations and Procedures International. 1992;**24**:33-37

[58] Ghabrial SS, Zaki MY. Dipolar cycloaddition reactions with quinazolinones: A new synthesis of azoloquinazolinone derivatives. The Journal of Chemical Research. 1997;**234**:1560

[59] Kubicová L, Kráľová K, Chobot V, Vytlačilová J, Jahodář L, Kaustová J. Synthesis and biological evaluation of quinazoline-4-thiones. Molecules. 2003;**8**:756-769

[60] Michael JP. Quinoline, quinazoline and acridone alkaloids. Natural Product Reports. 2005;**22**:627-646

[61] Cunningham D, Zalcberg J, Maroun J, James R, Clarke S, Maughan TS, et al. Efficacy, tolerability and management of raltitrexed (Tomudex™) monotherapy in patients with advanced colorectal cancer: A review of phase II/III trials. European Journal of Cancer. 2002;**38**:478-486

[62] Cao S-L, Feng Y-P, Jiang Y-Y. Synthesis and in vitro antitumor activity of 4(3H)-quinazolinone derivatives with dithiocarbamate side chains. Bioorganic and Medicinal Chemistry Letters. 2005;**15**:1915-1917

[63] Khalil AA, Abdel-Hamed SG, El-Subbagh HI. Substituted quinazolines, part 2. Synthesis and in-vitro anticancer evaluation of new 2-substituted mercapto-3H-quinazoline analogs. Archiv der Pharmazie und Berichte der Deutschen Pharmazeutischen Gesellschaft. 2003;**2**:95-103

[64] Van Zyl EF. A survey of reported synthesis of methaqualone and some positional and structural isomers. Forensic Science International. 2001;**122**:142-149

[65] El-Helby AA, Abdel-Wahab MHA. Design and synthesis of some new derivatives of 3H-quinazolin-4-one with promising anticonvulsant activity. Acta Pharmaceutica. 2003;**53**:127-138

[66] Gatadi S, Lakshmi TV, Nanduri S. 4(3H)-Quinazolinone derivatives: Promising antibacterial drug leads. European Journal of Medicinal Chemistry. 2019;**170**:157-172

[67] Sarma GV, Thomas J, Murugan V, Elango K. Nicotinyl incorporated quinazolinonyl thiadiazoles as possible NSAIDS. Indian Journal of Heterocyclic Chemistry. 1999;**9**:150-151

[68] Maarouf A, El-Bendary E, Goda FE. Synthesis and evaluation of some novel quinazolinone derivatives as diuretic agents. Archiv der Pharmazie. 2004;**337**:527-532

Chapter 5

Quinazolinone and Quinazoline Derivatives: Synthesis and Biological Application

Satyendra Mishra

Abstract

Drug discovery and optimization comprise one of the most significant targets in medicinal chemistry. Quinazoline and quinazolinone derivatives and nitrogen-containing heterocycles have received significant attention due to their widely and distinct biopharmaceutical activities. Quinazolines and quinazolinones are considered as noteworthy chemical for the synthesis of diverse physiological significance and pharmacological utilized molecules. Quinazolines are building blocks for about 150 naturally occurring alkaloids with a broad range of biological activity. The various substituted quinazolines and quinazolinones displayed important, for example, sedative hypnotics, antibacterial, anti-inflammatory, analgesic, antipsychotic, antifungal, antimalarial, anticonvulsant, anti-Parkinsonism, cancer, and other activities. This chapter aims to highlight the latest evidence of quinazolinone and quinazoline derivatives as a privileged scaffold in medicinal chemistry.

Keywords: quinazoline, quinazolinones, antioxidant and anticancer, antibacterial, structure-activity relationship

1. Introduction

Emergence of drug resistance has created a critical and unmet medical requirement for the innovation and development of novel classes of antibacterial agents [1–4]. Due to the appearance of drug resistance bacterial strains, there is an escalating need for the development of novel antibiotics to treat the resistant bacteria stain. Diverse set of biological activities of quinazolinones (fused heterocyclic system) such as anti-inflammatory, anticonvulsant, anticancer, antibacterial, antifungal, anti-HIV and anti-analgesic [5–16], have encouraged to abundant of medicinal chemists to investigate this fused heterocycles as a novel drug molecules. Several research groups have successfully investigated and reported the promising antimicrobial properties and structure-activity relationships (SAR) of various quinazolinone derivatives.

Quinazolines and quinazolinones emerged as a privileged class of nitrogen containing heterocyclic scaffolds; exhibits a broad spectrum of pharmacological activities, viz. anti-inflammatory, antitubercular, and antiviral activities [17]. Number of quinazoline derived compound have been approved as a drug; for example prazosin and doxazosine are used to treat benign prostatic hyperplasia and post-traumatic stress disorder [18], and erlotinib and gefitinib both are used for the curing of lung and pancreatic cancers (**Figure 1**) [19].

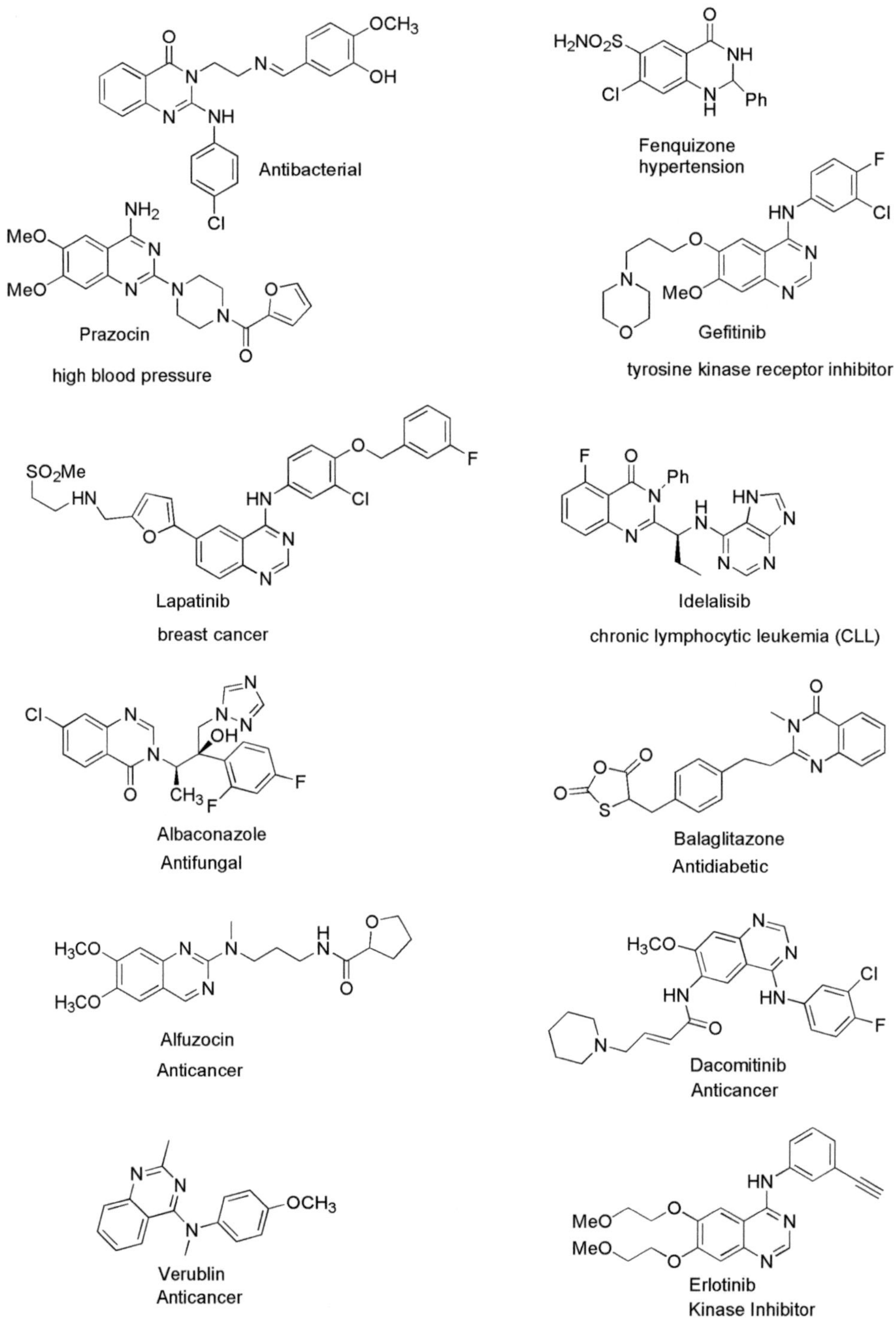

Figure 1.
Quinazoline and quinazolinone-based drugs.

Several quinazolinone-based drugs including idelalisib and fenquizone have been shown to exhibit a broad spectrum of antimicrobial, antitumor, antifungal, and cytotoxic activities [20]. Lapatinib has been displayed to be effective in combination therapy for breast cancer [21]. In the recent years, various synthetic strategies for the synthesis of quinazolines and quinazolinones derivatives have been developed to accomplish the budding requirements of medicinal chemist [22]. Many research groups have successfully utilized copper catalyzed Ullmann-type coupling procedures of aryl bromides and benzamidines for the synthesis of quinazoline derivatives [23].

2. Synthesis of quinazoline and quinazolinone derivatives

1. Synthesis of 4(3H)-quinazolinone using anthranilic acid or formyl anthranilamide [24].

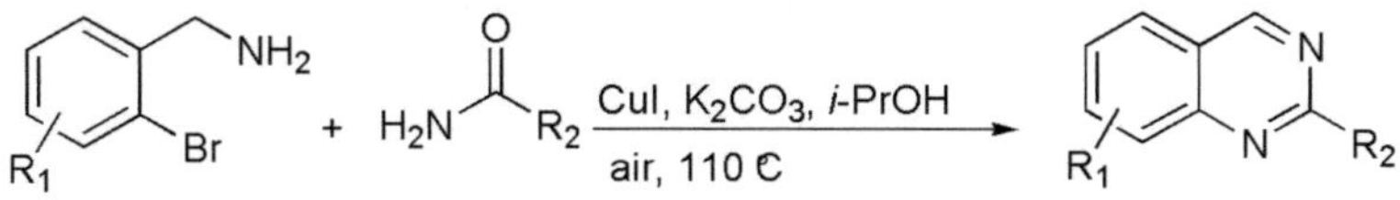

2. Via condensation reaction of 4-chloroanthranilic acid amide with triethyl orthoformate, the 7-chloro-substituted derivative has been prepared [25].

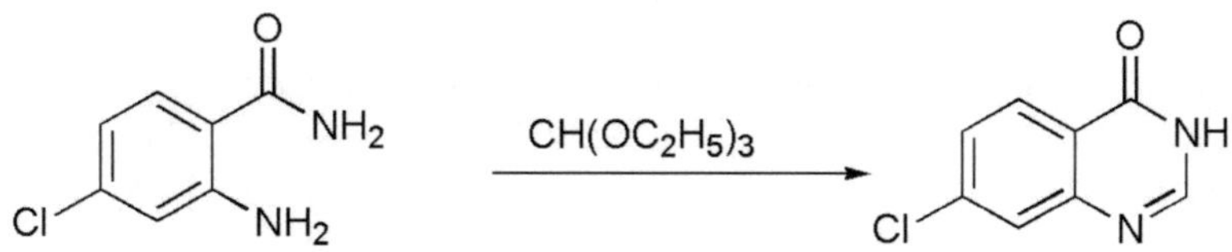

3. Quinazolin-4(3H)-one was synthesized by the reaction of anthranilic acid with excess formamide at 120°C in an open air. This is also known as Niementowski reaction [26].

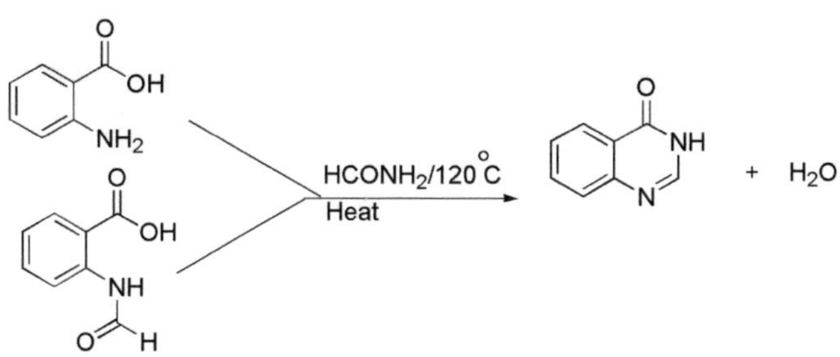

4. 2-styryl-4(3H)-quinazolinone derivatives were prepared using starting substrate 2-aminobenzonitrile with 3-phenyl cinnamoyl chloride. Under alkaline conditions, intramolecular cyclization of cinnamamide derivative was carried out to afford 2-styryl-4(3H)-quinazolinone. This procedure was tolerated to a wide range of different substituted benzene rings [27].

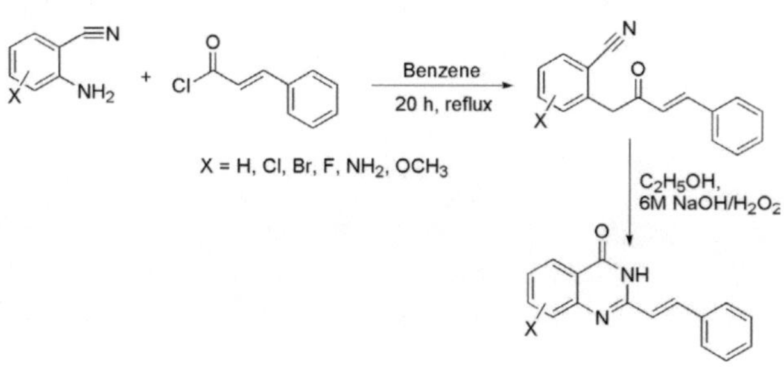

5. Reaction of anthranilic acid with ammonium acetate, followed by formamide under microwave at 200 W yields the desired 2-substituted-4(3H)-quinazolinones products [28].

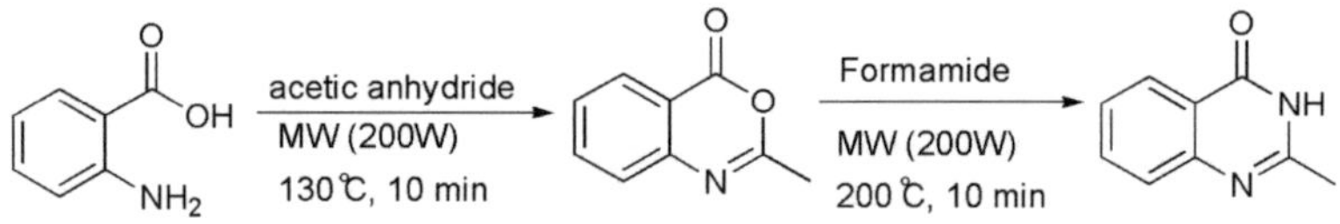

6. Reaction of anthranilamide with substituted aldehydes or ketones in 2,2,2-trifluoroethanol under reflux condition led to the formation of 2-substituted-2,3-dihydro-4(1H)-quinazolinones in excellent yields [27].

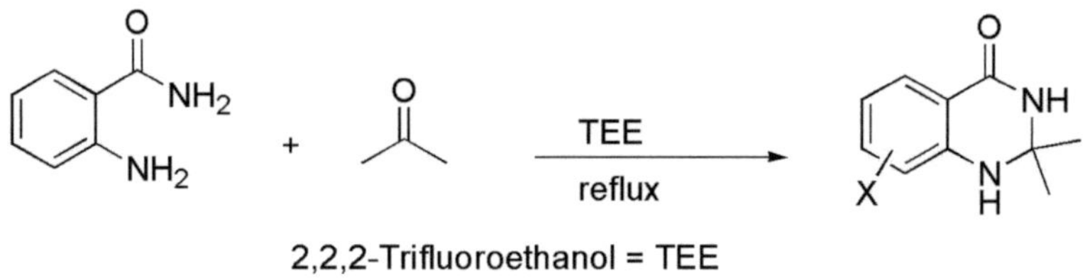

7. The amino-quinazolin-4(3H)-one was synthesized by means of the reaction of the corresponding methyl anthranilate with an excess amount of guanidine in ethyl alcohol containing sodium ethoxide in moderate yield [29].

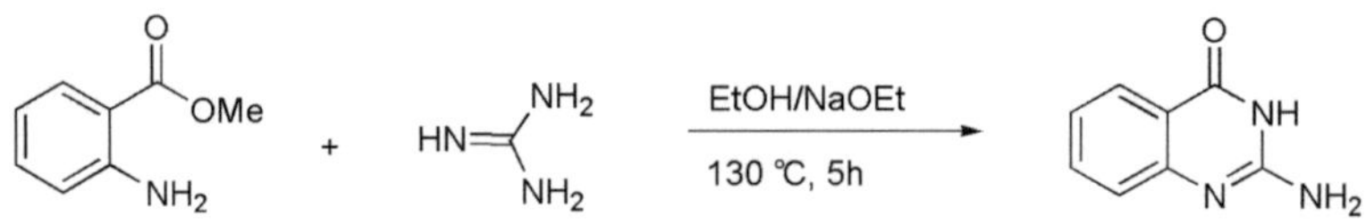

8. 4-Arylaminoquinazolines has vast biological potential as anticancer agents, thus there has been great interest in their syntheses. Through the reaction of 2-aminobenzonitrile with different substituted anilines and anhydrous aluminum chloride, amidines were readily produced. Highest yield of the amidine intermediates was obtained, when excess amounts of suitable aniline and aluminum chloride were used [30].

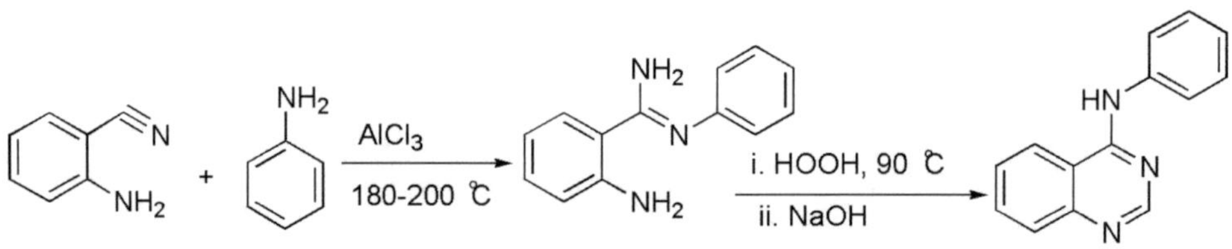

9. 2,3-disubstituted-4(3H)-quinazolinone derivatives were prepared through the treatment of N-acylanthranilic acid with the appropriate aryl amines in the presence of phosphorous oxychloride [31].

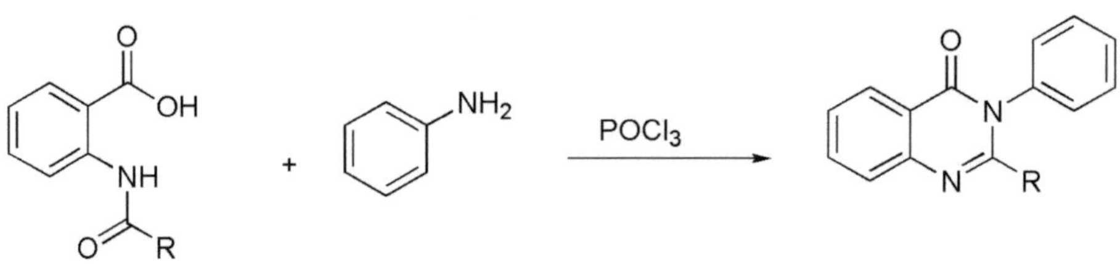

10. Benzoxazinone derivatives are the most widespread intermediates in the formation of 2,3-disubstituted quinazolinone derivatives. 2-methyl-4H-benzo[d][1,3]oxazin-4-one was prepared by refluxing mixture of anthranilic acid with acetic anhydride in acetic acid [32].

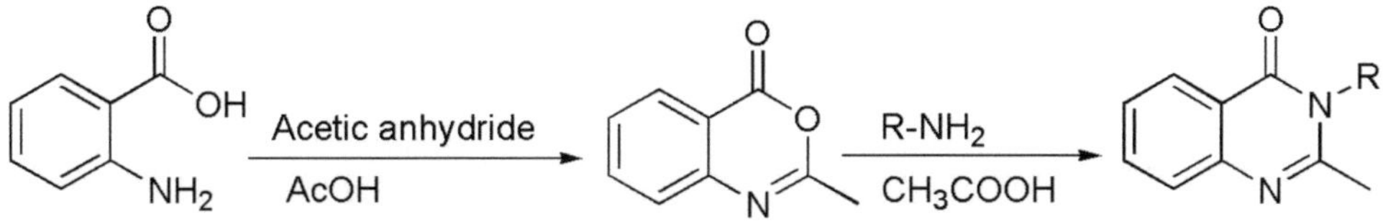

11. The reaction of 2-aminobenzonitrile with Grignard reagents yields the intermediates. The produced intermediate derivatives were very significant for getting many types of quinazoline derivatives. Upon their cyclization with acid chlorides, anhydrides, and formates, they formed the corresponding quinazoline derivatives in moderate to good yields. This general method for the preparation of various 2,4-disubstituted quinazoline derivatives is highly flexible and useful [33].

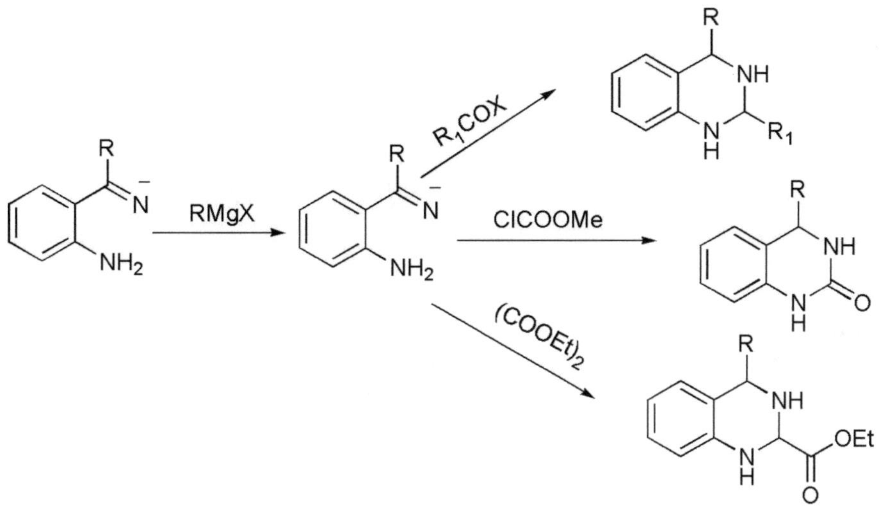

12. As shown in the scheme 2-chloromethyl-4-methyl-quinazoline derivatives were synthesized by the reaction 1-(2-amino-phenyl)-ethanone with HCl gas in anhydrous condition in presence of chloro acetonitrile to get 2-chloromethyl-4-methyl-quinazoline. Subsequently treatment of 2-chloromethyl-4-methyl-quinazoline with different amine derivative in presence of base furnished 2-chloromethyl-4-methyl-quinazoline derivatives [34].

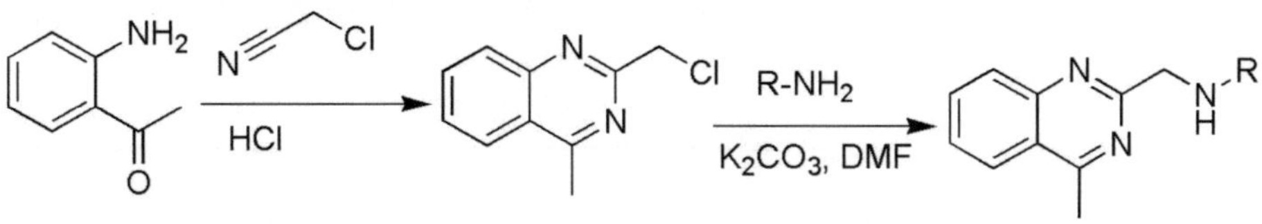

3. Biological activities of quinazolinone and quinazoline derivatives

Subsequently the innovation of quinazoline ring numeral of structural modifications have been made in order to raise the biological activities such as antitubercular,

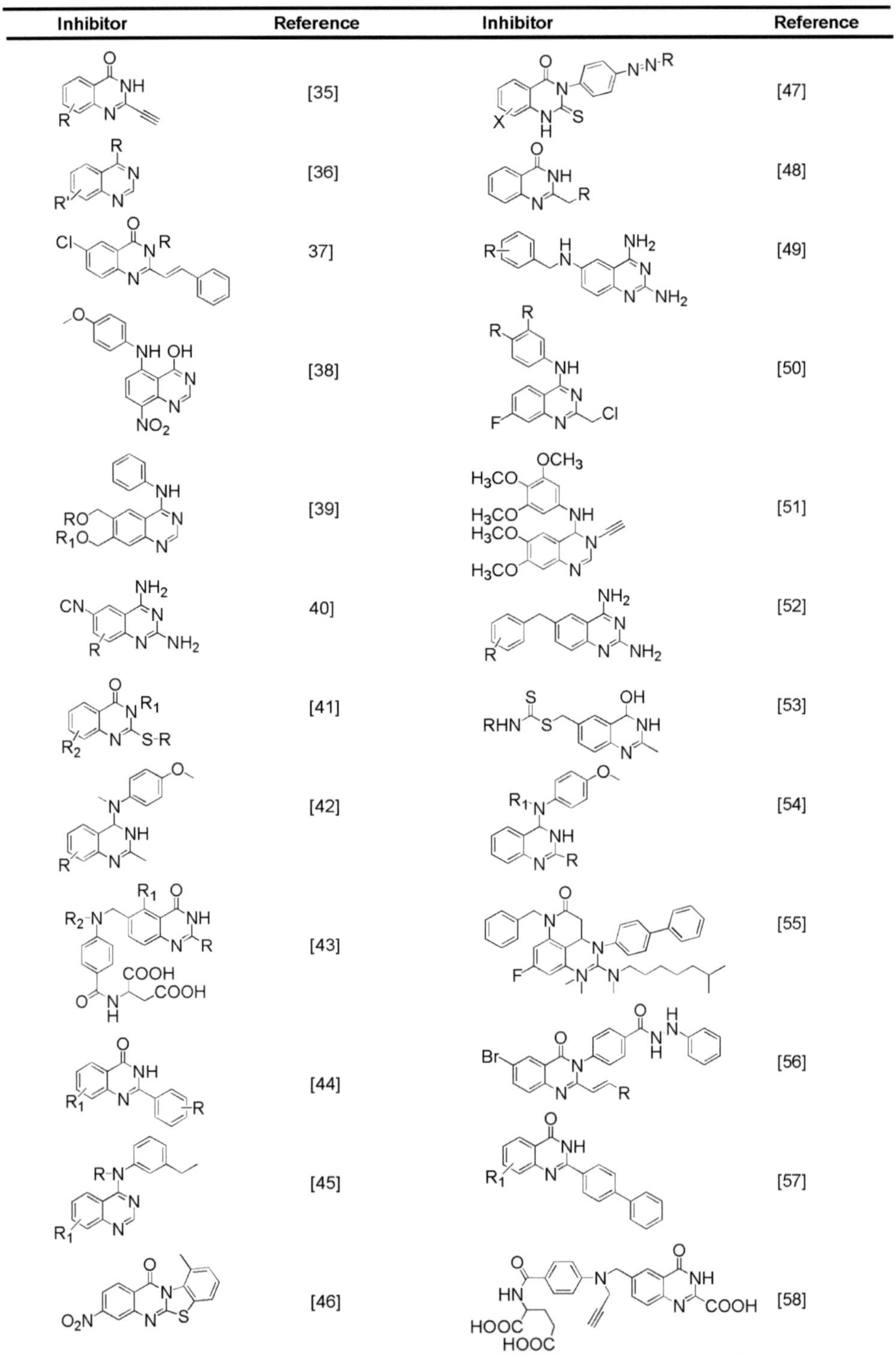

Figure 2.
Anticancer activities of quinazolinone and quinazoline derivatives.

antihistaminic, analgesic, anticonvulsant, antibacterial, antifungal, and anti-inflammatory activity which attracted the interest of medicinal chemists.

Cancerous augmentation is the main reasons of global human mortality. Numerous antineoplastic drugs are in the market and the majority of the compounds are under clinical trials. Studies make known that these antineoplastic drugs have exhibited the diverse kinds of side effects, as a result researchers around the world are engaged in the designing of more proficient and novel antineoplastic drugs. Recently, quinazoline and its derivatives have been considered as a novel class of neoplastic chemotherapeutic agents to facilitate activity against diverse tumors. Quinazoline is one of the most attractive novel bioactive compounds between all the heterocyclic compounds.

Quinazolinone derivatives, the privileged structures in the field of medicinal chemistry not only act as good anticancer agents but also act as good DNA intercalates [1, 2]. A systematic report is depicted herein for quinazoline ring. A number of quinazolinone and quinazoline derivatives (compounds **1–24**) have been reported for their various anticancer activities (**Figure 2**) [35–56].

A series of quinazolinone derived Schiff base derivatives were synthesized and evaluated for their in vitro H+/K+-ATPase inhibition. Many quinazolinone derived Schiff base exhibited outstanding potency, compared to the reference drug omeprazole. Especially, hydroxy and methoxy derivatives were the most potent compounds, contributing positively to gastric H+/K+-ATPase inhibition. Preliminary structure-activity relationship revealed that the compounds **25–30** with electron donating moiety (OH, OCH_3) were found to be excellent activity and compounds **31–34** with electron withdrawing moiety (Cl and NO_2) were found to be least antiulcer agents [57].

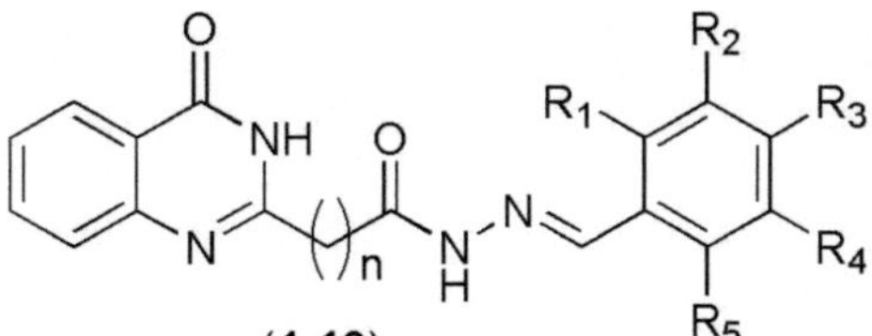

(1-10)

n = 2	n= 3	R_1	R_2	R_3	R_4	R_5
25	26	H	H	OH	H	H
27	28	H	H	OCH3	H	H
29	30	H	OCH3	OH	H	H
31	32	H	H	Cl	H	H
33	34	H	H	NO2	H	H

Quinazolinone derived Schiff base derivatives were also used as novel antioxidants and anti-inflammatory agents. The in vitro antioxidant activities of these compounds were evaluated and compared with commercial antioxidants viz. ascorbic acid (AA), gallic acid (GA), butylated hydroxytoluene (BHT), (DPPH) assay, etc. Data illustrates that quinazolinone derived Schiff base with electron donating moiety (OH, OCH_3) were found to be excellent antioxidants and compounds with electron withdrawing moiety (Cl, NO_2) were found to be excellent anti-inflammatory agents [58].

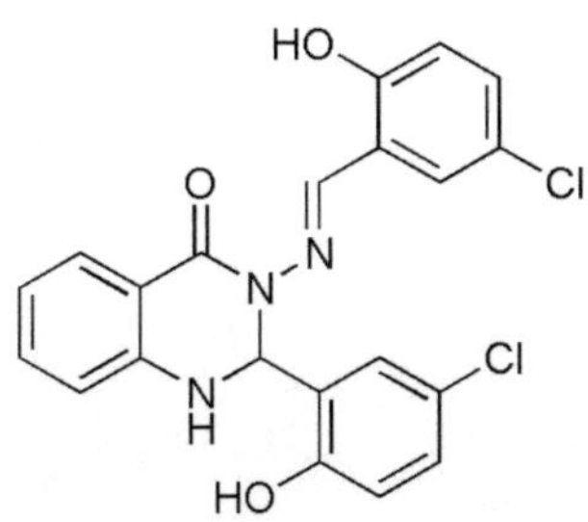

35

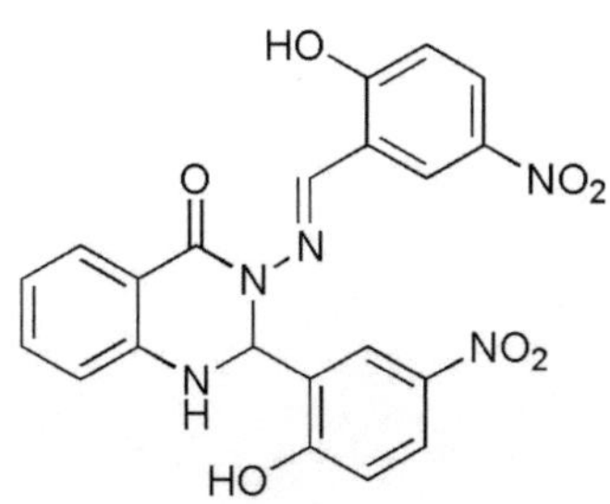

36

Plausible pathways induced by inhibitors were assessed by evaluating the cytotoxic effect of inhibitors such as 3-(5-chloro-2-hydroxybenzylideneamino)-2-(5-chloro-2-hydroxyphenyl)-2,3-dihydroquinazolin-41(H)-one (**35**) and 3-(5-nitro-2-hydroxybenzylideneamino)-2-(5-nitro-2-hydroxyphenyl)-2,3-dihydroquinazolin-4(1H)-one (**36**) on MCF-7, MDA-MB-231, MCF-10A and WRL-68 cells. MTT assay results of both the compounds showed significant inhibition of MCF-7 cell viability [59].

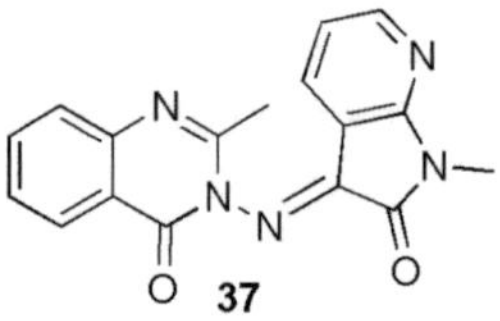

37

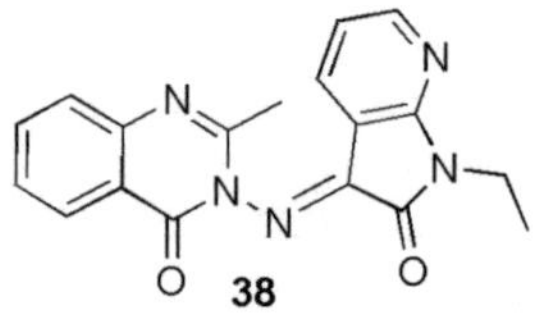

38

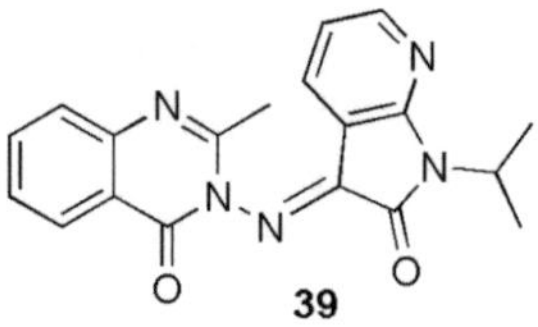

39

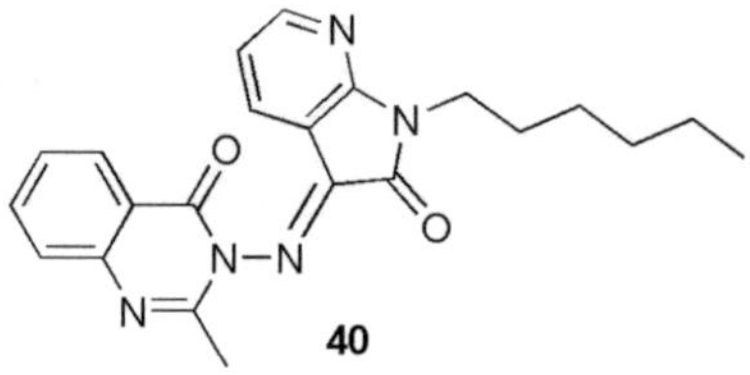

40

Azaisatins derivative containing 4(3H) quinazolinones has been designed and synthesized and were screened for their potential antimicrobial activities, which exhibited some authentic results towards testing organism *in vitro* and *in vivo* studies. Azaisatins derivatives with $-C_6H_{13}$ (**40**) display good antimicrobial activity compare to other synthesized Azaisatins [60].

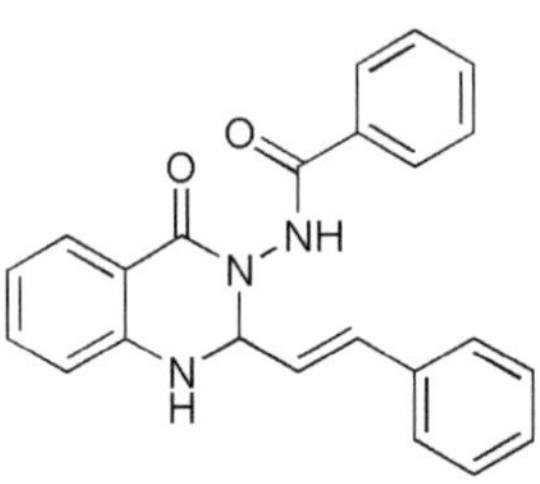

41

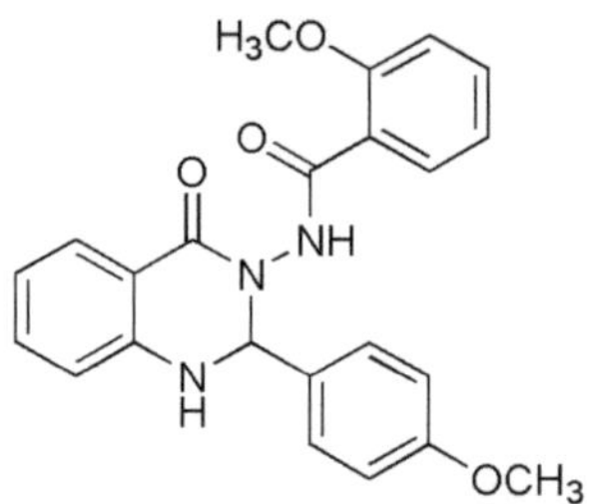

42

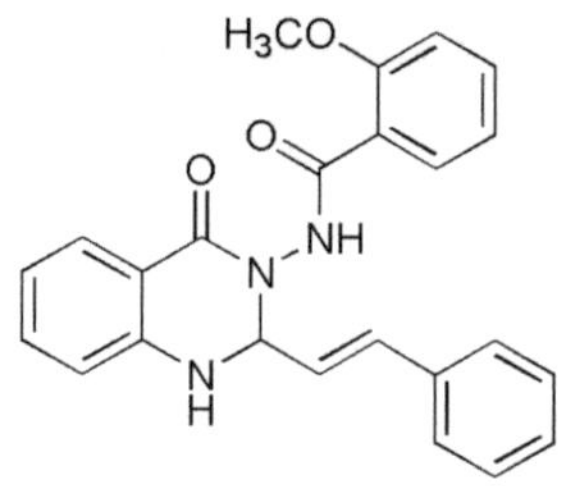

43

Quinazolinone derivatives containing 3-acrylamino motifs were screened for antifungal activities against four phytopathogenic fungi by minimum inhibitory concentration (MIC) method. Compounds **41–43**, exhibited broad antifungal activities and substituent's play important role in activities [61].

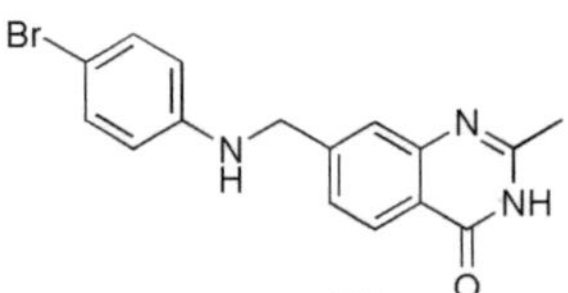

44

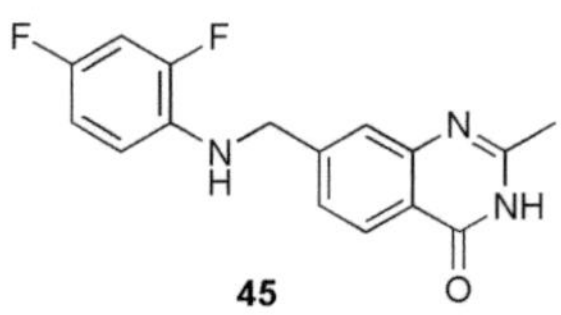

45

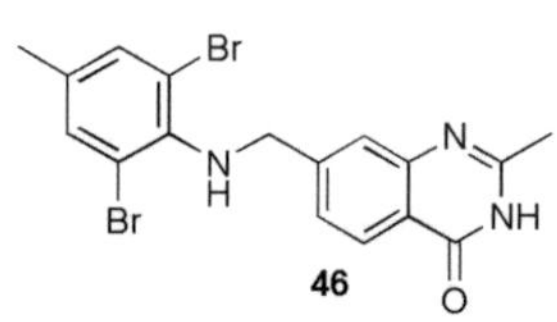

46

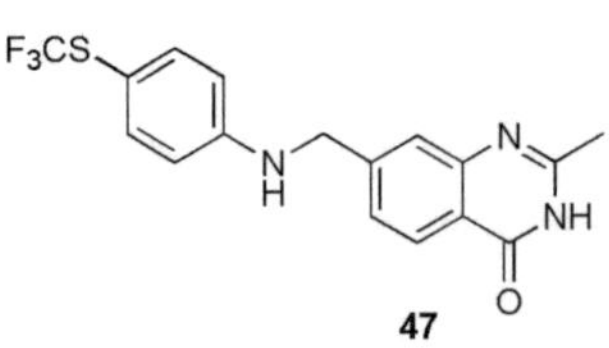

47

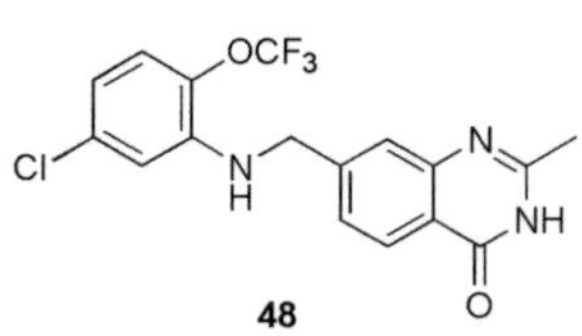

48

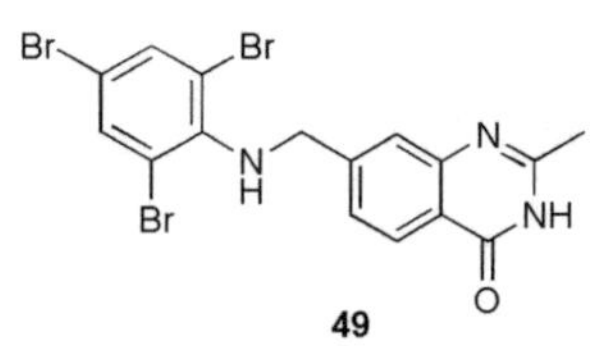

49

A series of novel quinazolinone derivatives containing an amino substituted amino moiety were reported for their cytotoxic and antibacterial activities. Among the synthesized compounds **47–49** showed broad-spectrum cytotoxic activities giants at least four cancer cell lines at low concentrations. Compounds **44–46** exhibited good to moderate antibacterial activities against gram positive and gram negative bacterial strains [62].

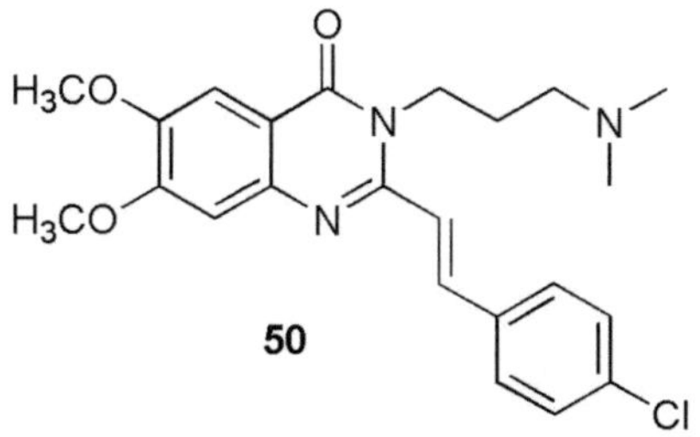

50

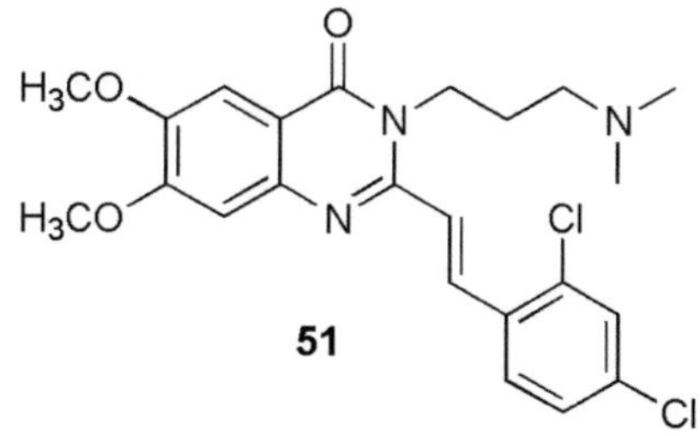

51

Quinazolinone derivatives manipulate mutant p53 proteins and their corresponding cellular response in p53 mutant cancer cells. Compounds **50** and **51** exhibited promising broad-spectrum anti-cancer effects, while **50** demonstrated selective and exclusive inhibition activity in p53 mutant cancer cell lines. Quinazolinone derivatives **50** dictate mutant p53 function for apoptotic cell death [63].

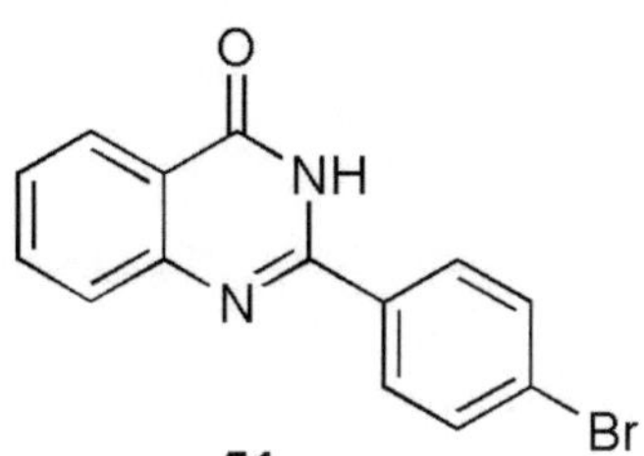

51

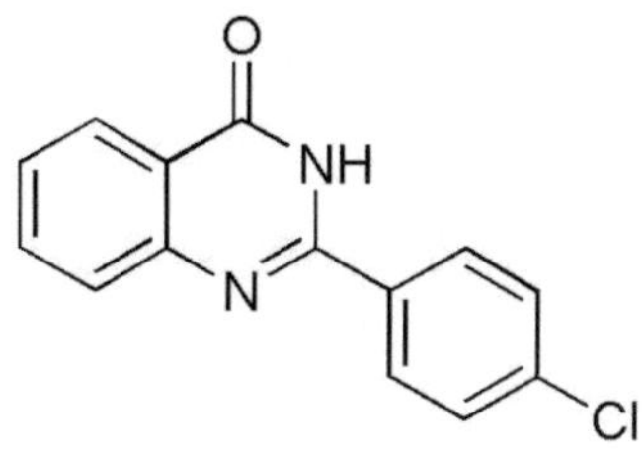

52

2-(4-bromophenyl)-quinazolin-4(3H)-one (**52A**) and 2-(4-chlorophenyl)-quinazolin-4(3H)-one (**52B**) exhibited α-glucosidase inhibitory activity with IC50 values of 12.5 ± 0.1 lM and 15.6 ± 0.2 lM, respectively. Spectroscopy methods were performed to analyze the inhibitory mechanisms of both compounds on α-glucosidase. The outcome of inhibitory mechanism disclosed, that the compounds, inhibited α-glucosidase in reversible and non-competitive manner. Briefly, the quinazolinone derivatives could be potentially promising candidates in the field of anti-diabetic agents development [64].

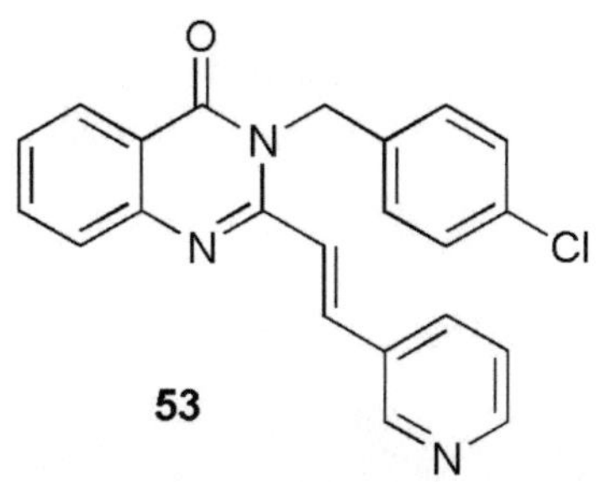

53

RAD51 is an essential component of the homologous recombination DNA repair pathway and is over expressed in drug-resistant cancers, including aggressive triple negative breast cancer (TNBC). Structure activity relationships study of quinazolinone derivatives showed that inhibitor (**53**) as a novel RAD51 inhibitor exhibited up to 15-fold enhanced inhibition of cell growth. Furthermore, inhibitors 17 notably hamper TNBC cell sensitivity to DNA damage. This would be potentially targeted therapy for cancer treatment [65].

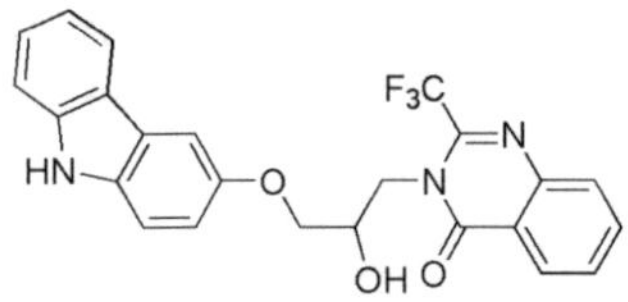

54

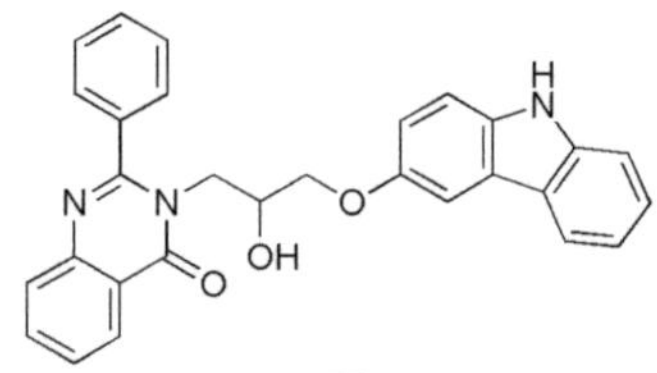

55

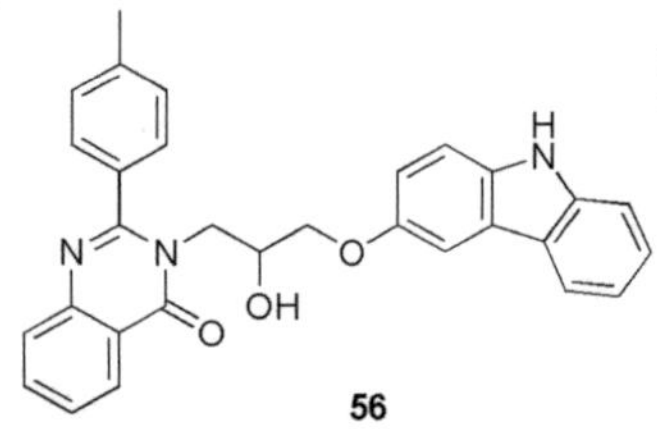

56

A series of novel carbazolyloxy phenylquinazoline derivatives have been developed as angiotensin converting enzyme (ACE) inhibitors. Amongst them compounds (**54**–**56**) showed maximum inhibitory potency in enzyme based assays. The most potent (**54**–**56**) compounds have common active site with the Lisinopril binding site [66].

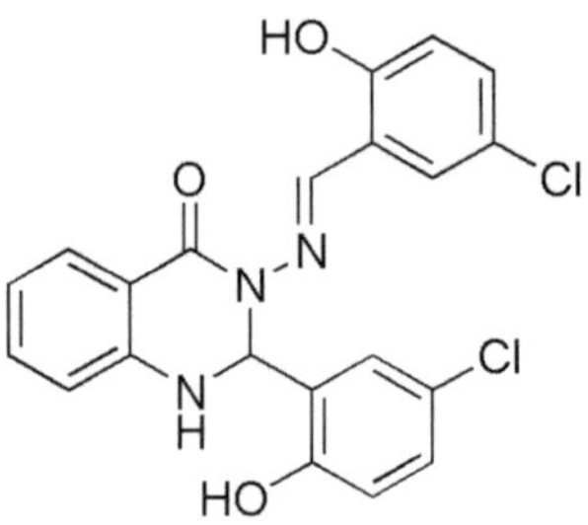

57

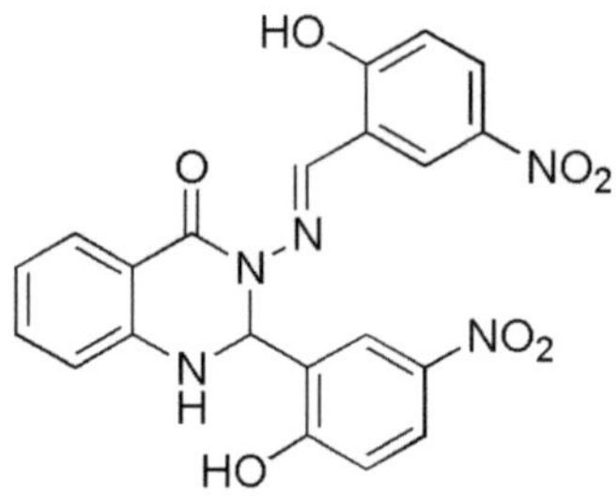

58

Compounds, 3-(5-chloro-2-hydroxybenzylideneamino)-2-(5-chloro-2-hydroxyphenyl)-2,3-dihydroquinazolin-41(H)-one (**57**) and 3-(5-nitro-2 -hydroxybenzylideneamino)-2-(5-nitro-2-hydroxyphenyl)-2,3-dihydroquinazolin-4(1H)-one (**58**) were screened for their cytotoxic effect on MCF-7, MDA-MB-231, MCF-10A and WRL-68 cells. The mechanism involved in apoptosis, induced by compound **57** and **58** was also evaluated. Additionally, caspase-8 illustrates significant potency, followed by inhibition of NF-κB activation in **57**- and **58**-treated MCF-7 cells. The results indicated that A and B could induce apoptosis via a mechanism that involves either extrinsic or intrinsic pathways [59].

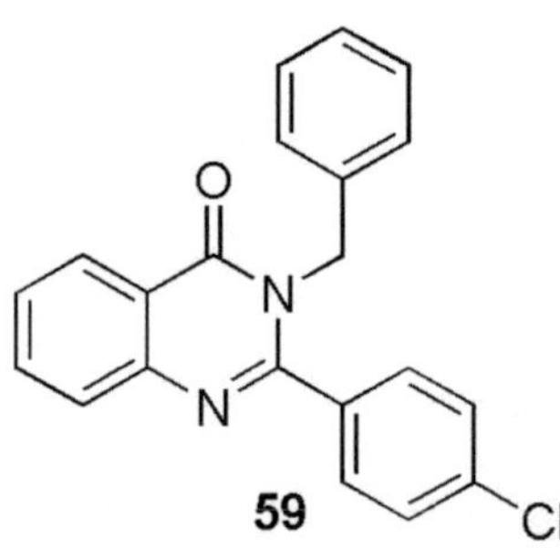

59

Substituted quinazolinones derivatives were tested for their antimicrobial activity against Gram-negative bacteria and Gram-positive bacteria. Among the prepared products, 3-benzyl-2-(4-chlorophenyl) quinazolin-4(3H)-one (**3a**) was found to exhibits the most potent *in vitro* anti-microbial activity against *Staphylococcus aureus*.

4. Conclusions

Over the past few decades, more effort has been established into searching of better drugs with minimal side effects. Herein number versatile synthetic procedures are discussed for the synthesis of quinazolinone and quinazoline derivatives. In general, quinazolinone and quinazoline derivatives are known to possess wide range of activities. A specific activity depends on the substituent present at an appropriate position of quinazoline. The study of natural and synthetic quinazolinone and quinazoline derivatives identified as potentially promising candidates for developing as novel therapeutic agents. There is possibility for further development as new research into study of medicinal chemistry related field.

Acknowledgements

Department of Science and Technology, India (DST-SERB/ECR/2015/000363) to SM.

Conflict of interest

The author declares no conflict of interest.

Author details

Satyendra Mishra
Department of Physical Sciences, Institute of Advanced Research, Gandhinagar, Gujarat, India

*Address all correspondence to: satyendramishra1@gmail.com

References

[1] Sugden R, Kelly R, Davies S. Combating antimicrobial resistance globally. Nature Microbiology. 2016;**1**:16187-16187

[2] Ventola CL. The antibiotic resistance crisis: Part 2: Management strategies and new agents. Pharmacology & Therapeutics. 2015;**40**:344-352

[3] Toner E, Adalja A, Gronvall GK, Cicero A, Inglesby TV. Antimicrobial resistance is aglobal health emergency. Health Security. 2015;**13**:153-155

[4] Bassetti M, Merelli M, Temperoni C, Astilean A. New antibiotics for bad bugs: Where arewe? Annals of Clinical Microbiology and Antimicrobials. 2013;**12**:22

[5] Khan I, Zaib S, Batool S, Abbas N, Ashraf Z, Iqbal J, et al. Quinazolines and quinazolinones as ubiquitous structural fragments in medicinal chemistry: An update on the development of synthetic methods and pharmacological diversification. Bioorganic & Medicinal Chemistry. 2016;**24**:2361-2381

[6] Abbas SY, El-Bayouki KA, Basyouni WM, Mostafa EA. New series of 4 (3*H*)-quinazolinone derivatives: Syntheses and evaluation of antitumor and antiviral activities. Medicinal Chemistry Research. 2018;**27**:571-582

[7] EI-Hashash MAEAM, Salem MS, Al-Mabrook SAM. Synthesis and anticancer activity ofnovel quinazolinone and benzamide derivatives. Research on Chemical Intermediates. 2018;**44**:2545-2559

[8] Noolvi MN, Patel HM, Bhardwaj V, Chauhan A. Synthesis and *in vitro* antitumor activity of substituted quinazoline and quinoxaline derivatives: Search for anticancer agent. European Journal of Medicinal Chemistry. 2011;**46**:2327-2346

[9] Dohle W, Jourdan FL, Menchon G, Prota AE, Foster PA, Mannion P, et al. Quinazolinone based anticancer agents: Synthesis, antiproliferative SAR, antitubulin activity, and tubulin co-crystal structure. Journal of Medicinal Chemistry. 2018;**61**:1031-1044

[10] Ighachane H, Sedra MH, Lazrek H. Synthesis and evaluation of antifungal activities of (3*H*)-quinazolin-4-one derivatives against tree plant fungi. Journal of Materials and Environmental Science. 2017;**8**:134-143

[11] Alaa AM, Abou-Zeid LA, ElTahir KEH, Ayyad RR, Magda AA, El-Azab AS. Synthesis, anti-inflammatory, analgesic, COX-1/2 inhibitory activities and molecular docking studies of substituted 2-mercapto-4 (3*H*)-quinazolinones. European Journal of Medicinal Chemistry. 2016;**121**:410-421

[12] Abuelizz HA, El-Dib RA, Marzouk M, Al-Salahi R. *In vitro* evaluation of new 2-phenoxy-benzo [g] [1, 2, 4] triazolo [1, 5-a] quinazoline derivatives as antimicrobial agents. Microbial Pathogenesis. 2018;**117**:60-67

[13] Yang L, Ge S, Huang J, Bao X. Synthesis of novel (*E*)-2-(4-(1*H*-1,2, 4-triazol-1-yl) styryl)-4-(alkyl/ arylmethyleneoxy) quinazoline derivatives as antimicrobial agents. Molecular Diversity. 2018;**22**:71-82

[14] Zhang L, Chen Q, Li XQ, Wu SQ, Wan JL, Ouyang GP. Synthesis and antibacterial activity of 2-substitued-(3-pyridyl)-quinazolinone derivatives. Journal of Heterocyclic Chemistry. 2018;**55**:743-749

[15] Al-Salahi R, Abuelizz HA, Ghabbour HA, El-Dib R, Marzouk M. Molecular docking study and antiviral evaluation of 2-thioxo-benzo[g] quinazolin-4(3H)-one derivatives. Chemistry Central Journal. 2016;**19**:10-21

[16] Alagarsamy V, Chitra K, Saravanan G, Raja Solomon V, Sulthana MT, Narendhar B. An overview of quinazolines: Pharmacological significance and recent developments. European Journal of Medicinal Chemistry. 2018;**151**:628-685

[17] Gupta T, Rohilla A, Pathak A, Akhtar MJ, Haider MR, Yar MS. Current perspectives on quinazolines with potent biological activities: A review. Synthetic Communications. 2018;**48**:1099-1127

[18] Akduman B, Crawford ED. Terazosin, doxazosin, and prazosin: Current clinical experience. Urology. 2001;**58**:49-54

[19] McConnell JD et al. The long-term effect of doxazosin, finasteride, and combination therapy on the clinical progression of benign prostatic hyperplasia. The New England Journal of Medicine. 2003;**349**:2387-2398

[20] Pao W, Miller VA. Epidermal growth factor receptor mutations, small-molecule kinase inhibitors, and non-small-cell lung cancer: Current knowledge and future directions. Journal of Clinical Oncology. 2005;**23**:2556-2568

[21] Johnston S, Pippen J Jr, Pivot X, Lichinitser M, Sadeghi S, Dieras V, et al. Lapatinib combined with letrozole versus letrozole and placebo as first-line therapy for postmenopausal hormone receptor-positive metastatic breast cancer. Journal of Clinical Oncology. 2009;**27**:5538-5546

[22] Kshirsagar UA. Recent developments in the chemistry of quinazolinone alkaloids. Organic & Biomolecular Chemistry. 2015;**13**:9336-9352

[23] Khan I, Ibrar A, Abbas N, Saeed A. Recent advances in the structural library of functionalized quinazoline and quinazolinone scaffolds. Synthetic approaches and multifarious applications. European Journal of Medicinal Chemistry. 2014;**76**:193-244

[24] Omar MA, Conrad J, Beifuss U. Copper-catalyzed domino reaction between 1-(2-halophenyl)methanamines and amidines or imidates for the synthesis of 2-substituted quinazolines. Tetrahedron. 2014;**70**:3061-3072

[25] MK MK, RL MK, Bost RW. 7-Chloro-4-(1-diethylamino-4-pentylamino)-quinazoline. Journal of the American Chemical Society. 1947;**69**:184-1184

[26] Alexandre F, Berecibar A, Wrigglesworth R, Besson T. Novel series of 8Hquinazolino[4,3-b]quinazolin-8-ones via two Niementowski condensations. Tetrahedron. 2003;**59**:1413-1419

[27] Qiao RZ, Xu BL, Wang YH. A facile synthesis of 2-substituted-2,3-dihydro-4(1H)-quinazolinones in 2,2,2-trifluoroethanol. Chinese Chemical Letters. 2007;**18**:656-658

[28] Nouira I, Kostakis IK, Dubouilh C, Chosson E, Iannelli M, Besson T. Decomposition of formamide assisted by microwaves, a tool for synthesis of nitrogen-containing heterocycles. Tetrahedron Letters. 2008;**49**:7033-7036

[29] Hess HJ, Cronin TH, Scriabine A. Antihypertensive 2-amino-4(3H)-quinazolinones. Journal of Medicinal Chemistry. 1968;**11**:130-136

[30] Szczepankiewicz W, Suwinski J, Bujok R. Synthesis of 4-arylaminoquinazolines and 2-aryl-4-arylaminoquinazolines from 2-aminobenzonitrile, anilines and formic acid or benzaldehydes. Tetrahedron. 2000;**56**:9343-9349

[31] Ager IR, Harrison DR, Kennewell PD, Taylor JB. Synthesis and central nervous system activity of quinazolones related to 2-methyl-3-(o-tolyl)-4(3H)-quinazolone (methaqualone). Journal of Medicinal Chemistry. 1977;**20**:379-386

[32] Welch WM, Ewing FE, Huang J, Menniti FS, Pagnozzi MJ, Kelly K, et al. Atropisomeric quinazolin-4-one derivatives are potent noncompetitive amino-3-hydroxy-5-methyl-4-isoxazolepropionic acid (AMPA) receptor antagonists. Bioorganic & Medicinal Chemistry Letters. 2001;**11**:177-181

[33] Bergman J, Brynolf A, Elman B, Vuorinen E. Synthesis of quinazolines. Tetrahedron. 1986;**42**:3697-3706

[34] Dempcy J, Searle MS, Maynard AJ, Williams HE. Design of 2-ethenyl and 2-(2′-haloethyl) substituted quinazolinones (3) as anticancer agents. Bioorganic & Molecular Chemistry. 1992;7:60

[35] Arnold D, Hassan SY, Khattab SN, Bekhit AA, Amer A. A series of novel 4,5,6,7-tetra substituted quinazolines and screened for anticancer activity. Bioorganic & Medicinal Chemistry Letters. 1995;**16**:1753

[36] Raffa D, Elder MC, Daidone G, Maggio B, Merickech M, Plescia S, et al. Synthesis, cytotoxicity and inhibitory effects on tubulin polymerization of a new 3-heterocyclo substituted 2-styryl quinazolinones. European Journal of Medicinal Chemistry. 2004;**39**:299-304

[37] Tian W, Qin L, Song Q, He L, Ai M, Jin Y, et al. A novel synthetic analog of 5,8-disubstituted quinazolines blocks mitosis and induces apoptosis of tumor cells by inhibiting microtubule polymerization. PLoS ONE. 2010;**5**(5):10499-10506

[38] Fathima ASE, Awadallah FM, Ibrahin NA, Said EG, Kamel G. Synthesis of novel 4-aryl amino-6,7-disubstituted quinazolin derivatives as antitumor activity. European Journal of Medicinal Chemistry. 1999;**53**:141

[39] Yan SJ, Zheng H, Huang C, Yan YY, Lin J. Synthesis of highly functionalized 2,4-diaminoquinazolines as anticancer and anti HIV agents. Bioorganic & Medicinal Chemistry Letters. 2010;**20**(15):4432-4435

[40] Al-Rashood ST, Aboldahb IA, Nagi MN, Abdel-Aziz AAM, Abdel-Hamide SG, Yousef KM, et al. Synthesis, dihydrofolate reductase inhibition, antitumor testing, and molecular model study of some new 4(3H)-quinazolinone analogs. Bioorganic & Medicinal Chemistry Letters. 2006;**14**:8608-8621

[41] Forsch RA, Wright JE, Rosowsky A. Synthesis and in-vitro antitumor activity of thiophene analogues of 5-chloro-5,8-dideazafolic acid. Bioorganic & Medicinal Chemistry. 2002;**10**:2067-2076

[42] Sirisoma N, Pervin A, Zhang H, Jiang S, Adam Willardsen J, Anderson MB, et al. Discovery of N-methyl-4-(4-methoxyanilino)-quinazolines as potent apoptosis inducers: Structure activity relationship of the quinazoline ring. Bioorganic & Medicinal Chemistry Letters. 2010;**20**(7):2330-2334

[43] Hour MJ, Huang LJ, Kuo SC, Xia Y, Bastow K, Nakanishi Y, et al. 6-Alkylamino and 2,3-dihydro-3′-methoxy-2-phenyl-4-quinazolinones and related compounds: Their synthesis, cytotoxicity, and inhibition of tubulin polymerization. Journal of Medicinal Chemistry. 2000;**43**(23):4479-4487

[44] Lehner RS, Norris T, Santafianos DP. Synthesis and anticancer activity of 4-(3-ethylphenylamino)-quinazolines. Chemical Abstracts. 2000;**133**(20):769

[45] Sherbeny EL, Magda A. Synthesis of certain pyrimido-[2,1-b] benzothiazole and benzothiazolo[2,3-b]-quinazoline derivatives for *in-vitro* antitumor and antiviral activities. Arzneimittel Forschung. 2000;**50**(II):848-853

[46] El-Hiti GA, Abdel-Megeed MF, Zied TMM. Synthesis and reaction of

some 3-aryl-2-thioxo quinazolin-4(3H)-ones. Indian Journal of Chemistry. 2002;**41B**:1519-1522

[47] Spirkova K, Stankovsky S, Mrvova A, Cipak L. Synthesis and biological activity of some 2-substituted quinazolin-4-ones. Chemical Papers. 1999;**53**(4):272-275

[48] Mang J, Mielcke TR, Mascarello A, Filippi-Chiela E, Zanin RF, Lenz G. Synthesis and anticancer activities of some new 2,4-diamino-6-substituted quinazoline derivatives. European Journal of Medicinal Chemistry. 1994;**48**:255

[49] Li HZ, He HY, Han YY, Gu X, He L, Qi QR, et al. A general synthetic procedure for 2-chloromethyl-4(3H)-quinazolinone derivatives and their utilization in the preparation of novel anticancer agents with 4-anilinoquinazoline scaffolds. Molecules. 2010;**15**(12):9473-9485

[50] Wang YD, Miller K, Boschelli DH, Ye F, Wu B, Floyd MB, et al. Inhibitors of tyrosine kinase: The preparation and structure activity relationship of 4-anilino-3-cyanoquinolines and 4-anilino quinazolines. Bioorganic & Medicinal Chemistry Letters. 2000;**10**(21):2477-2480

[51] Papoulis AT, Rosowsky A, Forsch RA, Queener SF. Synthesis, anti-parasitic, and antitumor activity of 2,4-diamino-6-(arylmethyl)-5,6,7, 8-tetrahydroquinazoline analogues of piritrexim. Journal of Medicinal Chemistry. 1999;**42**(6):1007-1017

[52] Cao SL, Feng YP, Jiang YY, Liu SY, Ding GY, Li RT. Synthesis and in-vitro antitumor activity of 4(3H)-quinazolinone derivatives with dithiocarbamate side chains. Bioorganic & Medicinal Chemistry Letters. 2005;**15**:1915-1917

[53] Sirisoma N, Pervin A, Zhang H, Jiang S, Willardsen JA, Anderson MB, et al. Discovery of N-(4-methoxyphenyl)-N-2-dimethylquinazolin-4-amine: A potent apoptosis inducer and efficacious anticancer agent with high blood brain barrier penetration. Journal of Medicinal Chemistry. 2009;**52**(8):2341-2351

[54] Jung SY, Lee SH, Kang HB, Park HA, Chang SK, Kim J, et al. Antitumor activity of 3,4-dihydroquinazoline dihydrochloride in A549 xenograft nude mice. Bioorganic & Medicinal Chemistry Letters. 2010;**20**(22):6633-6636

[55] Murgan V, Thomas CC, Rama Sarma GVS, Kumar EP. Synthesis of 2-substituted quinazolin-4(3H)-ones as a new class of anticancer agents. Indian Journal of Pharmaceutical Sciences. 2003;**65**(4):386-389

[56] Chinigo GM, Paige M, Grindrod S, Hamel E, Dakshanamurthy S, mChruszcz M, et al. Asymmetric synthesis of 2,3-dihydro-2-arylquinazolin-4-ones: Methodology and application to a potent fluorescent tubulin inhibitor with anticancer activity. Journal of Medicinal Chemistry. 2008;**51**(15):4620-4631

[57] Rakesh KP, Shantharam CS, Manukumar HM. Synthesis and SAR studies of potent H+/K+-ATPase inhibitors of quinazolinone-Schiff's base analogues. Bioorganic Chemistry. 2016;**68**:1-8

[58] Rakesh KP, Manukumar HM, Channe Gowda D. Schiff's bases of quinazolinone derivatives: Synthesis and SAR studies of a novel series of potential anti-inflammatory and antioxidants. Bioorganic and Medicinal Chemistry. 2015;**25**(5):1072-1077

[59] Zahedifard M, Faraj FL, et al. Synthesis, characterization and apoptotic activity of quinazolinone Schiff base derivatives toward MCF-7 cells via intrinsic and extrinsic apoptosis pathways. Scientific Reports. 2015;**5**:11544. DOI: 10.1038/srep11544

[60] Devi KA, Sarangapani M, Sriram. Synthesis and antimicrobial activity of some quinazolinones derivatives. International Journal of Drug Development and Research. 2012;**4**(3):324-327

[61] Zhang J, Liu J, Ma Y, Ren D, Cheng P, Zhao J, et al. One-pot synthesis and antifungal activity against plant pathogens of quinazolinone derivatives containing an amide moiety. Bioorganic & Medicinal Chemistry Letters. 2016;**6**(9):2273-2277

[62] Zhan X, Xu Y, Qi Q, Wang Y, Shi H, Mao Z. Synthesis, cytotoxic, and antibacterial evaluation of quinazolinone derivatives with substituted amino moiety. Biodiversitas. 2018;**15**(3):e1700513

[63] Zhang G-H, Xue W-B, An Y-F, Yuan J-M, Qin J-K, Pan C-X, et al. Distinct novel quinazolinone exhibits selective inhibition in MGC-803 cancer cells by dictating mutant p 53 function. European Journal of Medicinal Chemistry. 2015;**95**:377-387

[64] Wei M, Chai WM, Wang R, Yang Q, Deng Z, Peng Y. Quinazolinone derivatives: Synthesis and comparison of inhibitory mechanisms on a-glucosidase. Bioorganic and Medicinal Chemistry. 2017;**25**:1303-1308

[65] Ward A, Dong L, Harris JM, Khanna KK, Al-Ejeh F, Fairlie DP, et al. Quinazolinone derivatives as inhibitors of homologous recombinase RAD51. Bioorganic & Medicinal Chemistry Letters. 2017;**27**(14):3096-3100

[66] Venkatesh R, Kasaboina S, Gaikwad HK, Janardhan S, Bantu R, Nagarapu L, et al. Design and synthesis of 3-(3-((9H-carbazol-4-yl)oxy)-2-hydroxypropyl)-2-phenylquinazolin-4(3H)-one derivatives to induce ACE inhibitory activity. European Journal of Medicinal Chemistry. 2015;**96**:22-29

Chapter 6

Synthesis and Pharmacological Research Regarding New Compounds with Quinazolin-4-One Structure

Cornel Chiriță, Carmen Limban, Diana Camelia Nuță, Emil Ştefănescu, Simona Negreş, Cristina Elena Zbârcea, Cristina Daniela Marineci, Oana Cristina Șeremet, Mihaela Trandafir (Bratu), Alexandru Vasile Missir and Ileana Cornelia Chiriță

Abstract

The quinazoline scaffold is found in the chemical structure of many marketed drugs used in CNS disorders as antidepressants, anxiolytics, or hypnotics. Also, the carbamate ester derivatives have different certain therapeutic actions, such as hypnotic or parasympathomimetic ones. We have obtained new 4(3H)-quinazolinones by bringing together in the same structure the quinazoline nucleus and carbamate ester group. The compounds named Q1–Q5 were characterized by measuring the melting points, by determining the infrared and NMR spectra, and by elemental analysis. The pharmacological tests evidenced that the compounds have a very low acute toxicity, lethal doses being >2000 mg/kg bw. The compounds had different actions observed in forced swimming test (FST), tail suspension test (TST), or elevated plus maze (EPM), probably influenced by the presence of different radicals on the nucleus. Thus, Q1 with a nitro group in structure manifested the highest antidepressant effect, showing a reduction of immobilization time in FST and TST. On the other hand, Q3 and Q5, with two groups methoxy, respective ethoxy, had a slight anxiolytic effect, highlighted by an increase of the time spent in open arms and a decrease of the time spent in closed arms of EPM.

Keywords: quinazolines, synthesis, antidepressant, anxiolytic, forced swimming test, tail suspension test, elevated plus maze

1. Introduction

The quinazolines constitute an important class of fused heterocycles that are also known as 5,6-benzopyrimidine or benzo[a]pyrimidine, benzo-1,3-diazine, or 1,3-diazanaphthalene. The name quinazoline was first proposed for this compound by Weddige, due to the similarity with cinnoline and quinoxaline [1, 2]. The 4-hydroxyquinazolines, tautomeric with 4-keto-3,4-dihydroquinazolines, are

 IntechOpen

commonly named 4(3H)-quinazolones; they are an important class of heterocyclic compounds, more than 200 natural compounds having this basic structure [3].

The stability of the quinazolinone nucleus was an important reason why many drug chemistry studies followed synthesis in this class of compounds; a large number of compounds have been synthesized and evaluated for their different biological activities. The first renowned quinazoline marketed drug was methaqualone, used for its sedative-hypnotic effects since 1951 [4].

The quinazoline scaffold is found in the chemical structure of many marketed drugs used in CNS disorders, having antidepressant, hypnotic, and sedative effect (afloqualone, diproqualone, etaqualone, and methaqualone), or used as anticonvulsant (piriqualone), antipyretic, nonsteroidal anti-inflammatory (fluproquazone and proquazone), and antidiabetic agents (balaglitazone, raltitrexed, ispinesib, and halofuginone) [5].

The anticonvulsant action of these compounds has become a priority for pharmacological testing [2], many of these studies highlighting the importance of methyl group at the second position of quinazolin-4(3H)-one [6]. As Gatadi et al. have also shown, more and more recent studies are concerned with investigating the antimicrobial potential of quinazolone derivatives because bacterial strains have developed resistance to available chemotherapeutics [7].

Several research groups have successfully investigated and reported the promising antimicrobial properties and structure-activity relationships (SAR) of various 4(3H)-quinazolinone derivatives [8, 9].

As Hieu et al. reported in recent studies, novel hydroxamic acids incorporating quinazoline-4(3H)-one are a promising class of molecules of interest for the treatment of cancer [10, 11].

On the other hand, it is known that the carbamate ester derivatives have different certain therapeutic actions; this class includes physostigmine, neostigmine, pyridostigmine, rivastigmine, methocarbamol, and carisoprodol.

Carbamates are also of interest for their action as HIV-1 protease inhibitors (darunavir, amprenavir, and atazanavir) [12].

We have concentrated our research activity on bringing together the quinazoline nucleus and carbamate ester group in the same 4(3H)-quinazolinone structure [13].

2. Synthesis of 4(3H)-quinazolinone derivatives

We obtained the new 4(3H)-quinazolone derivatives using the acylation of potassium 2-[2-methyl-3-(4-oxoquinazolin-3(4H)-yl)-aceto]-hydroxamate with aromatic acid chlorides in the presence of dioxane [14]. The general reaction scheme is presented (**Figure 1**).

We obtained the new derivatives by applying the following working technique:

Around 0.68 g (0.0025 mol) of potassium 2-[2-methyl-3-(4-oxoquinazolin-3(4H)-yl)-aceto]-hydroxamate was heating, and then 0.0025 moles of acid chloride in 20 mL of dioxane was gradually added; a white precipitate (potassium chloride)

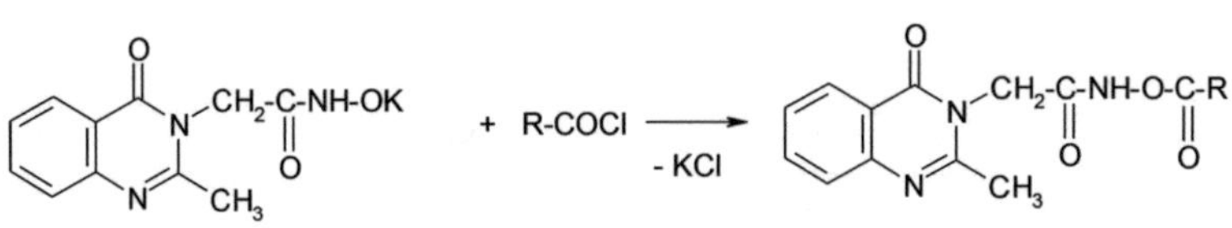

R= 4-NO_2, 3-CH_3, 3,5-$(OCH_3)_2$, 3,4,5-$(OCH_3)_3$, 3,5-$(OC_2H_5)_2$

Figure 1.
The synthesis of new 4(3H)-quinazolinone derivatives.

is formed. The reaction mixture is refluxed for 3 hours and then filtered. The filtrate was evaporated to dryness by mild heating under vacuum to give the crude product. The new compounds are recrystallized from isopropanol.

All chemicals and solvents were supplied by Sigma-Aldrich Chemical Company. All the solvents were distilled and dried before use.

Melting points were measured in open capillary tubes on an Electrothermal 9100 apparatus, and they are uncorrected.

Infrared spectra were recorded on a FT/IR-solid in ATR spectrometer (the signal intensities (height) were denoted by the following abbreviations: w = weak, m = medium, s = strong, v = variable).

The NMR spectra were recorded on a Varian 2000 and Bruker Fourier 300 instruments at room temperature, operating at 300 MHz for 1H and 75 MHz for 13C. The chemical shifts were recorded in δ units (ppm), relative to residual peak of the deuterated dimethyl sulfoxide (DMSO-d6). Tetramethylsilane (TMS) was used as internal standard. The coupling constants values are reported in hertz, and the splitting patterns are abbreviated as follows: s = singlet, d = doublet, t = triplet, m = multiplet, and b = broad.

The elemental analyses were performed on a PerkinElmer CHNS/O Analyzer Series II 2400 apparatus, and the results were in agreement with the calculated values.

For a better interpretation of spectral data, we used the numbering of the atoms presented in **Table 1**.

Compound Q1: 3-(4-Nitro-phenyl-carbonyl-oxi-carbamoyl-methyl) -2-methylquinazolin-4(3H)-one.

$C_{18}H_{14}N_4O_6$ (Mr 382.33); m.p. 183–184°C; yield 38.7%.

Elemental analysis: Calculated: C 56.55%, H 3.69%, N 14.65%; found: C 56.78%, H 3.56%, N 14.84%.

^{1}H-NMR (dmso-d6, δ ppm, J Hz, T = 308 K): 12.84(s, 1H, H-12); 8.39 (d, 2H, H-16, H-18, J = 9.4); 8.27(d, 2H, H-15, H-16, J = 9.2); 8.11(dd, 1H, H-5, J = 7.8, J = 1.6); 7.82(td, 1H, H-7, J = 7.8, J = 1.6); 7.61(bd, 1H, H-8, J = 7.8); 7.51 (td, 1H, H-6, J = 7.8, J = 1.4); 4.98(s, 2H, H-10); 2.58(s, 3H, H-9).

^{13}C-NMR (dmso-d6, δ ppm, T = 308 K): 164.99(CO-11); 162.69(CO-13); 161.13(C-4); 155.11(C-2); 150.87(C-17); 146.97(C-1a); 131.86(C-14); 119.57(C-4a); 134.71(C-7); 131.13(C-16, C-18); 126.60(C-8); 126.56(C-6); 126.28(C-5); 124.26(C-15, C-19); 43.96(C-10); 22.73(C-9).

FT-IR (solid in ATR, ν cm^{-1}): 3164 m; 3077w; 3017w; 2976 m; 1783s; 1711 m; 1642vs; 1597vs; 1528vs; 1473 m; 1416w; 1386w; 1347s; 1233 m; 1066 m; 975 m; 873w; 844w; 774 m; 710 m; 658w.

Compound Q2: 3-(3-Methyl-phenyl-carbonyl-oxi-carbamoyl-methyl) -2-methylquinazolin-4(3H)-one.

$C_{19}H_{17}N_3O_4$ (Mr 351.36); m.p. 151–52°C, yield 42.5%.

Elemental analysis: Calculated: C 64.95%, H 4.88%, N 11.96%; found 64.80%, H 4.85%, N 12.04%.

^{1}H-NMR (dmso-d6, δ ppm, J Hz, T = 308 K): 12.61(s, 1H, H-12); 8.11(dd, 1H, H-5, 1.4, 8.0); 7.85(t, 1H, H-15,); 7.82(td, H-7, 8.0, 1.4); 7.85÷7.80(m, 2H, H-15, H-19); 7.62(dd, 1H, H-8, 1.4, 8.0); 7.56(bd, 1H, H-17, 7.7); 7.51(td, 1H, H-6, 8.0, 1.4); 7.47(t, 1H, H-18, 7.7); 4.97(s, 2H, H-10); 2.58(s, 3H, H-9); 2.38(s, 3H, H-16′).

^{13}C-NMR (dmso-d6, δ ppm, T = 308 K): 164.86(CO-11); 164.14(CO-13); 161.14(C-4); 155.17(C-2); 147.00(C-1a); 138.78(C-16); 126.42(C-14); 119.61(C-4a);135.18(C-19); 134.70(C-7); 129.85(C-15); 129.10(C-17); 126.73(C-18); 126.60(C-8); 126.55(C-6); 126.30(C-5); 43.94(C-10); 22.73(C-9); 20.73(C-16′).

FT-IR (solid in ATR, ν cm^{-1}): 3200 m; 3004w; 2956w; 1774 m; 1672vs; 1599s; 1519w; 1468 m; 1385 m; 1340w; 1264 m; 1170 m; 1065 m; 972 m; 860w; 778 m; 733 m; 695w.

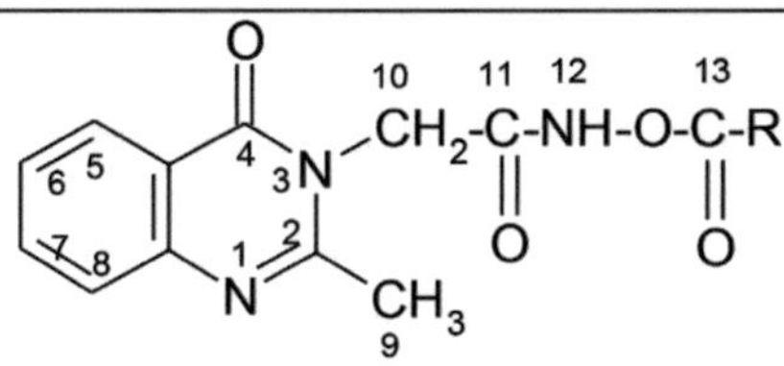

No	R	No	R
1.	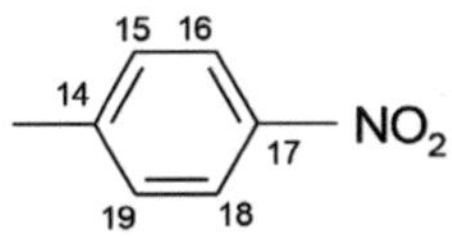	4.	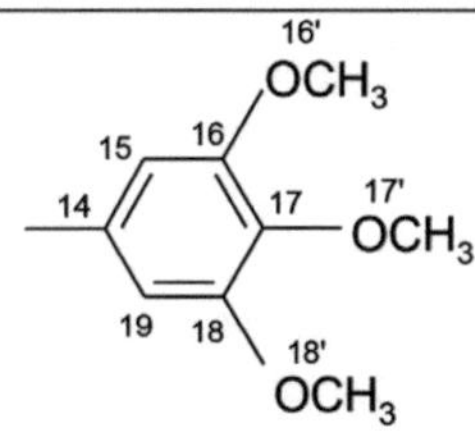
2.	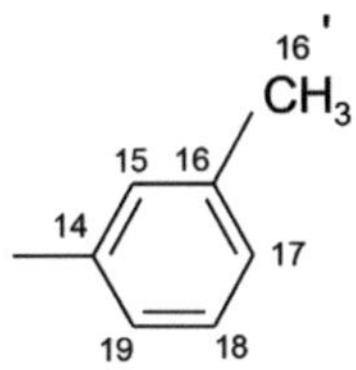	5.	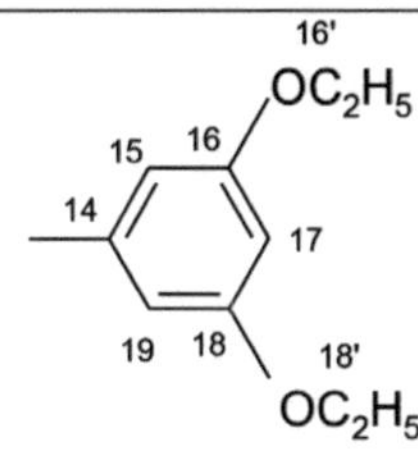
3.	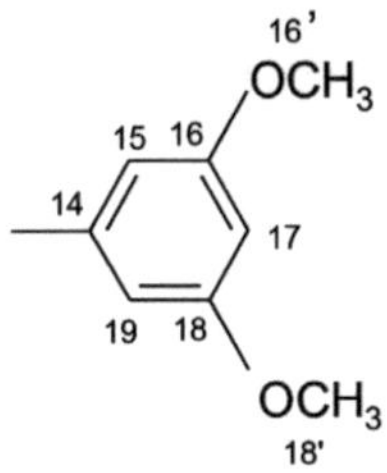		

Table 1.
Numbering of atoms for spectral interpretation.

Compound Q3: 3-(3,5-Dimethoxy-phenyl-carbonyl-oxi-carbamoyl-methyl)-2-methylquinazolin-4(3H)-one.

$C_{20}H_{19}N_3O_6$ (Mr 397.39); m.p. 148–49°C, yield 48.1%.

Elemental analysis: Calculated: C 60.45%, H 4.82%, N 10.57%; found C 60.60%, H 4.96%, N 10.65%.

^{1}H-NMR (dmso-d6, δ ppm, J Hz, T = 308 K): 12.58(s, 1H, H-12); 8.10(dd, 1H, H-5, J = 7.8 Hz, J = 1.6 Hz); 7.79(td, 1H, H-7, J = 7.8, J = 1.6); 7.62(bd, 1H, H-8, J = 7.8); 7.51(td, 1H, H-6, J = 7.8, J = 1.5); 7.11(d, 2H, H-15, H-19, J = 2.3); 6.87 (t, 1H, H-17, J = 2.3); 4.96(s, 2H, H-10); 3.81(s, 6H, H-16′, H-18′); 2.58(s, 3H, H-9).

^{13}C-NMR (dmso-d6, δ ppm, T = 308 K): 164.86(CO-11); 163.74(CO-13); 161.13(C-4); 160.69(C-16, C-18); 155.09(C-2); 147.05(C-1a); 128.33(C-14); 119.61(C-4a); 134.65(C-7); 126.63(C-8); 126.51(C-6); 126.26(C-5); 106.99(C-19, C-15); 106.46(C-17); 55.66(C-16′, C-18′); 43.93(C-10); 22.73(C-9).

FT-IR (solid in ATR, ν cm^{-1}): 3239w; 3092w; 2949w; 2844w; 1775 m; 1676vs; 1600vs; 1500w; 1469 m; 1431 m; 1390 m; 1352 m; 1305 m; 1212 m; 1195 m; 1179 m; 1165 m; 1081w; 1051 m; 1016 m; 976 m; 931w; 877w; 848w; 772 m; 747w.

Compound Q4: 3-(3,4,5-Trimethoxy-phenyl-carbonyl-oxi-carbamoyl-methyl)-2-methylquinazolin-4(3H)-one.

$C_{21}H_{21}N_3O_7$ (Mr 427.41); m.p.152–53°C, yield 45.3%.

Elemental analysis: Calculated: C 59.01%, H 4.95%, N 9.83%; found C 58.92%, H 5.05%, N 10.02%.

^{1}**H-NMR** (dmso-d6, δ ppm, J Hz, T = 308 K): 12.58(s, 1H, H-12); 8.11(dd, 1H, H-5, J = 7.3, J = 1.6); 7.82(td, 1H, H-7, J = 7.3, J = 1.6); 7.62(bd, 1H, H-8, J = 7.3); 7.51(td, 1H, H-6, J = 7.3, J = 1.4); 7.29(s, 2H, H-15, H-19); 4.96(s, 2H, H-10); 3.85 (s, 6H, H-16′, H-18′); 3.76(s, 3H, H-17′); 2.58(s, 3H, H-9).

^{13}**C-NMR** (dmso-d6, δ ppm, T = 308 K): 164.88(CO-11); 163.64(CO-13); 161.13(C-4); 155.10(C-1a); 152.99(C-16, C-18); 147.05(C-17); 142.69(C-14); 121.30(C-14); 119.61(C-4a); 134.65(C-7); 126.63(C-8); 126.51(C-6); 126.26 (C-5); 106.83(C-15, C-19); 60.25(C-17′); 56.13(C-16′, C-18′); 43.90(C-10); 22.72(C-9).

FT-IR (solid in ATR, ν cm^{-1}): 3200 w; 2979 w; 1766 m; 1680 vs; 1601 s; 1512 m; 1465 m; 1416 wm; 1405 w; 1277 m; 1250 m; 1209 m; 1188 m; 1144 m; 1071 m; 1017 m; 972 m; 870 m; 772 m.

Compound Q5: 3-(3,5-Diethoxy-phenyl-carbonyl-oxi-carbamoyl-methyl)-2-methylquinazolin-4(3H)-one.

$C_{22}H_{23}N_3O_6$ (Mr 425.44); m.p. 158–59°C, yield 35.8%.

Elemental analysis: Calculated: C 62.11%, H 5.45%, N 9.88%; found C 62.25%, H 5.58%, N 10.05%; m.p. 158–159°C; yield 35.8%.

^{1}**H-NMR** (dmso-d6, δ ppm, J Hz, T = 308 K): 12.59(s, 1H, H-12); 8.11(dd, 1H, H-5, J = 7.8, J = 1.6); 7.82(td, 1H, H-7, J = 7.8, J = 1.6); 7.62(bd, 1H, H-8, J = 7.8); 7.51(td, 1H, H-6, J = 7.8, J = 1.4); 7.07(d, 2H, H-15, H-19, J = 2.1); 6.82(t, 1H, H-17, J = 2.1); 4.96(s, 2H, H-10); 4.07(q, 4H(CH_2), H-16′, H-18′, J = 6.9); 2.58(s, 3H, H-9); 1.32(t, 6H(CH_3), H-16′, H-18′, J = 6.9).

^{13}**C-NMR** (dmso-d6, δ ppm, T = 308 K): 163.78(CO-11); 163.14(CO-13); 161.09(C-4); 159.88(C-16, C-18); 155.03(C-2); 147.01(C-1a); 128.25(C-14); 119.57(C-4a).

134.58(C-7); 126.58(C-8); 126.44(C-6); 126.20(C-5); 107.30(C-16, C-18); 107.21(C-17); 63.64(CH_2-16′, CH_2-18′); 43.89(C-10); 22.67(C-9); 14.43(CH_3-16′', CH_3-18′).

FT-IR (solid in ATR, ν cm^{-1}): 3219 w; 2980 w; 2937 w; 2882 w; 1778 m; 1675 vs; 1602 s; 1510 w; 1471 w; 1451 m; 1389 m; 1372 m; 1355 m; 1301 m; 1179 vs; 1116 m; 1086 m; 1058 s; 976 m; 934 m; 859 m; 829 w; 776 m; 747 m; 708 w; 692 w; 676 w; 658 m.

3. Pharmacological research on new 4(3H)-quinazolinone derivatives

3.1 Objective

The main objective of our study was to assess the potential pharmacological actions on central nervous system of five new 4(3H)-quinazolinone derivatives. For this purpose, we evaluated first the acute oral toxicity on mice, using the "up and down" method, in accordance with European Guidelines regarding the ethic of experimental research on animals [15]. These guides mention that the substances expected to have a low degree of toxicity can be tested using the limit test at 2000 mg/kg bw and only in special situations at the dose of 5000 mg/kg bw. After assessing the toxicity level of the substances, we performed a battery of tests to highlight the pharmacological potential of the new compounds on the central nervous system. Thus, we determined the antidepressant effect using the forced swimming test (FST) and tail suspension test (TST); the effect on anxiety using the elevated plus maze (EPM), Ugo Basile, Italy; and the effect on the motor activity with the activity cage (Ugo Basile).

All pharmacological tests were performed on mice, following all the existing protocols from the Laboratory of Pharmacology, Faculty of Pharmacy, UMF "Carol Davila" Bucharest.

3.2 Materials and methods

We used for the pharmacological experiments 85 male, white, NMRI mice, weighing 26 ± 1.7 g. The animals were supplied by the rodent farm of the University of Medicine and Pharmacy "Carol Davila" Bucharest. The animals were housed in ventilated cages with free access to food and water. The temperature and the relative humidity were kept constant (22–24°C, 45–60%).

All experimental procedures were carried out in accordance with the Directive 2010/63/UE of 22 September 2010, regarding the protection of animals used for experimental and other scientific purposes. All experimental procedures were approved by the Ethical Committee of the Faculty of Pharmacy, Bucharest. The experiment was conducted in May 2018.

For acute toxicity evaluation, we used five groups of three mice each, which received the new five quinazolinone derivatives in the dose of 2000 mg/kg bw p.o. All animals were followed for 14 days regarding any sign of lethality, body weight evolution, or behavior changes.

The pharmacological tests on central nervous system were performed on 70 mice, which were initially subjected to the Ugo Basile activity cage test. The parameter on which the animals were divided into groups was the horizontal motor activity, measured for every period of 5 min. The mice were divided into 7 groups of 10 individuals each, with similar average responses and standard deviations between groups.

The compounds were administered as shown below:

- Group I (control)—distilled water 0.1 ml/10 g orally
- Group II (reference)—amitriptyline 10 mg/kg bw susp. 0.1% orally
- Group III—Q1 100 mg/kg bw, susp. 1% orally
- Group IV—Q2 100 mg/kg bw, susp. 1% orally
- Group V—Q3 100 mg/kg bw, susp. 1% orally
- Group VI—Q4 100 mg/kg bw, susp. 1% orally
- Group VII—Q5 100 mg/kg bw, susp. 1% orally

We decided to test the animals at a dose of 100 mg/kg bw, considering the level of 1/20 of the dose administered in acute toxicity test which provided no lethal effects to mice.

The pharmacological tests were performed as follows:

- After 1 day of administration: activity cage and FST
- After 14 days of administration: activity cage and TST
- After 15 days of administration: EPM

FST was chosen after acute administration because it proved effective and consistent in testing antidepressant effect after one single dose [16].

All tests were conducted respecting the following protocol: in the testing chamber, the animals were kept in artificial light. Each individual was administered with a 7-min delay from the previous one (5 min for the test itself and 2 min to clean the device before testing the next animal) so that all of them could be tested after the same time interval from the moment of receiving the treatment.

Determination of motor activity assessed the influence of new compounds on mice motility and desire to explore. The duration of this test in the activity cage was 5 min for each mouse. The animals were placed each time in the same corner of the device [17].

Determination of immobility time of mice in forced swimming test (FST), was originally described by Porsolt [16, 18]. Each mouse was placed into a glass cylinder (25 cm height, 30 cm diameter) containing water at a temperature of 23 ± 1°C. The test duration was 6 min, the first two for accommodation and the next four for the actual determination of the immobilization time. The mouse is considered immobilized when it ceases to struggle and remains in an immobile, characteristic position, with minimal movements for keeping the head above the water.

Determination of immobility time of mice in tail suspension test (TST) involves the same principle as FST, the difference being the nature of the inducing factor of the depressive state, the suspension of the animal by the tail. In this test, there is no need for accommodation, so the determination of the immobility time starts from the beginning of the experiment [19].

Determination of anxiolytic potential of the compounds was performed using the *elevated plus maze Ugo Basile*. The mouse was placed in the center of the device, and it was left free to explore the maze. We determined the time spent into the open arms and the time spent into the closed arms. We considered that the mouse was in one of the arms when all four limbs were in that arm [20].

3.2.1 Statistical analysis

Statistical calculation used the software GraphPad Prism version 8.0.0 for Windows, GraphPad Software, San Diego, California, USA, www.graphpad.com. Statistical comparison between groups used the ANOVA test. In case it indicated a statistical significance, the Tukey posttest was performed comparing all groups 2 × 2. When appropriate, we determined the Pearson correlation coefficient.

Normality of response distribution in collectivity was tested with D'Agostino and Pearson test.

3.3 Results and discussion

Acute toxicity research did not lead to any lethality for the five new compounds administered at a dose of 2000 mg/kg bw. According to the "up and down" method, higher doses are not recommended, and the compounds can be classified in the

	Group I	Group II	Group III	Group IV	Group V	Group VI	Group VII
M	623.2	622.1	623.5	622.6	623.0	622.8	623.9
SD	100.7	104.1	86.1	107.7	95.25	99.29	107.52

M = average, SD = standard deviation.

Table 2.
The horizontal motor activity (HMA) of the groups formed after the selection process.

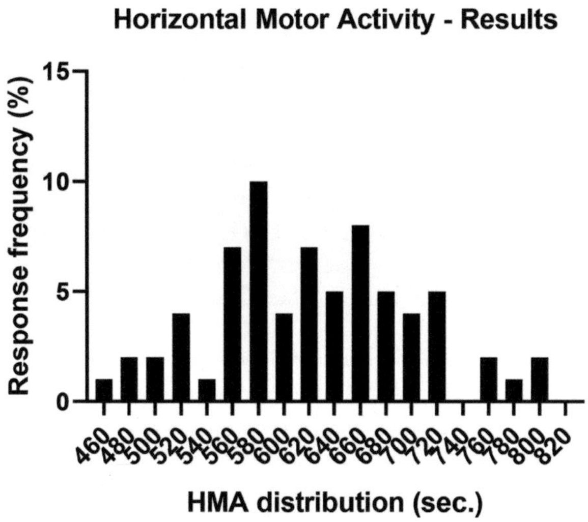

Figure 2.
Normal distribution of initial HMA values.

category "very low toxicity." Evolution of body weight, determined every other day for 14 days, was similar between treated mice and control ones. Changes in weight were small, being statistically insignificant. Motor behavior was similar, and response to auditory and tactile stimuli was present. No animal showed any palpebral ptosis, and the appearance of the fur and tail remained unchanged during the experiment.

After the *initial determination of motor activity*, the mice were divided into seven homogenous groups, each containing 10 individuals. Their baseline mean horizontal motor activity is shown in **Table 2**, and the Gaussian distribution of the individual results is highlighted in **Figure 2**:

After acute administration, compound Q5 had the most intense effect on motor activity, with a 23.69% reduction in HMA compared with the control group. This result was statistically significant according to ANOVA followed by Tukey posttest. The same compound reduced VMA with 37.39% compared with the control, but this result was not significant. Q3 was another compound which reduced the motor activity by more than 20%, but the results were not significant. The other three new quinazolinones had limited influence on motor activity. Amitriptyline used as the reference substance did not influence significantly the motor activity after one administration, as it can be observed in **Tables 3** and **4**.

The immobilization time in FST after acute administration was influenced differently by the new five compounds and was correlated with the results obtained in motor activity testing. In **Table 5**, it can be seen that compounds that have decreased the most intense motor activity (Q3, Q5) have led to an increase in immobilization time, with 44.09 and 41.24% compared with the reference group. Compound Q1 had the most intense antidepressant effect after one dose, quantified by reducing the immobilization time with 20.39% compared with the control group.

	M ± SD	ANOVA	Effect % vs. control	Tukey posttest/ control	Effect % vs. reference	Tukey posttest/ reference
Group I (control)	500.50 ± 91.25	<0.001**	—	—	—	—
Group II (reference)	477.50 ± 80.27		−4.59%	ns	—	—
Group III (Q1)	499.00 ± 46.82		−0.29%	ns	−4.50%	ns
Group IV (Q2)	472.50 ± 57.33		−5.59%	ns	−1.05%	ns
Group V (Q3)	398.50 ± 52.26		−20.37%	ns	−16.54%	ns
Group VI (Q4)	436.50 ± 107.1		−12.78%	ns	−8.58%	ns
Group VII (Q5)	381.90 ± 107.0		−23.69%	<0.05*	−20.02%	ns

**Statistical significance*
***High statistical significance*

Table 3.
The horizontal motor activity (HMA) of the groups after one administration.

	M ± SD	ANOVA	Effect % vs. control	Effect % vs. reference
Group I (control)	59.90 ± 26.94	ns	—	—
Group II (reference)	50.90 ± 15.59		−15.02%	—
Group III (Q1)	50.70 ± 14.21		−15.35%	−0.39%
Group IV (Q2)	52.50 ± 19.73		−12.35%	−3.14%
Group V (Q3)	45.50 ± 15.38		−24.04%	−10.60%
Group VI (Q4)	46.30 ± 17.01		−22.70%	−9.03%
Group VII (Q5)	37.50 ± 28.16		−37.39%	−26.33%

Table 4.
The vertical motor activity (VMA) of the groups after one administration.

	M ± SD	ANOVA	Effect % vs. control	Tukey posttest/ control	Effect % vs. reference	Tukey posttest/ reference
Group I (control)	121.10 ± 34.51	<0.0001***	—	—	—	—
Group II (reference)	98.20 ± 18.97		−18.91%	ns	—	—
Group III (Q1)	96.40 ± 14.47		−20.39%	ns	−1.83%	ns
Group IV (Q2)	101.30 ± 16.40		−16.35%	ns	3.15%	ns
Group V (Q3)	138.70 ± 28.13		14.53%	ns	41.24%	<0.01**
Group VI (Q4)	132.70 ± 26.03		9.57%	ns	35.13%	<0.05*
Group VII (Q5)	141.50 ± 27.73		16.84%	ns	44.09%	<0.01**

**Statistical significance.*
***High statistical significance.*
****Statistical significance. Very high statistical significance.*

Table 5.
The immobilization time in FST after one administration.

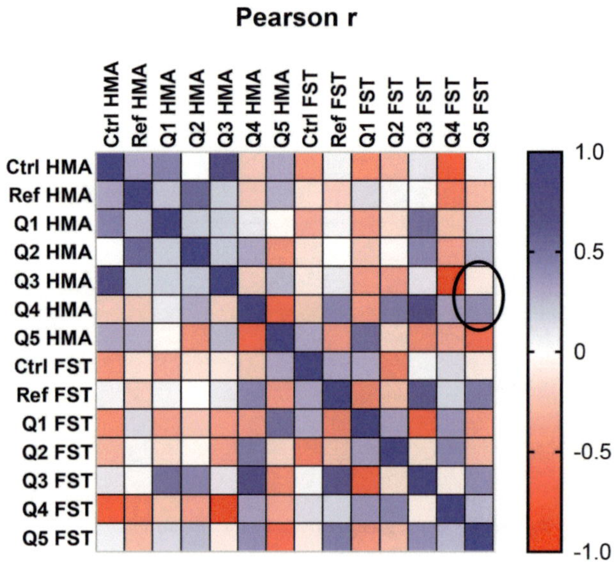

Figure 3.
Pearson correlation between data obtained in activity cage test and forced swimming test after acute administration (black circle—the best correlation for a compound between immobilization time in FST and horizontal movements activity).

	M ± SD	ANOVA	Effect % vs. control	Tukey posttest/ control	Effect % vs. reference	Tukey posttest/ reference
Group I (control)	471.50 ± 72.75	<0.0001***	—	—	—	—
Group II (reference)	382.90 ± 64.70		−18.79%	<0.05*	—	—
Group III (Q1)	479.50 ± 65.91		1.69%	ns	25.22%	<0.05*
Group IV (Q2)	482.50 ± 72.71		2.33%	ns	26.01%	<0.05*
Group V (Q3)	404.30 ± 50.62		−14.25%	ns	5.58%	ns
Group VI (Q4)	468.30 ± 56.68		−0.67%	ns	22.30%	ns
Group VII (Q5)	365.50 ± 52.63		−22.48%	<0.01**	−4.54%	ns

**Statistical significance.*
***High statistical significance.*
****Very high statistical significance.*

Table 6.
The horizontal motor activity (HMA) of the groups after 14 days of administration.

It was interesting to find the degree of Pearson correlation between the values obtained in the two tests after acute administration. As it can be observed in **Figure 3**, the best correlation of data is between values of group Q5—HMA and FST—with a coefficient of −0.622.

After 2 weeks of daily administration, the motor activity of the mice has illustrated a significant reduction of HMA in Q5 group, with 22.48% compared with control, which can be seen in **Table 6**. Also, in accordance with the known fact that amitriptyline has a sedative pharmacological profile, it determined a marked decrease of horizontal movements, with 18.79% compared with control.

Regarding the vertical movements, amitriptyline and the new compounds Q3 and Q5 produced the largest decrease, with effects between 14.63 and 34.03%, but ANOVA showed no statistical significance (**Table 7**).

	M ± SD	ANOVA	Effect % vs. control	Effect % vs. reference
Group I (control)	56.70 ± 17.13	ns	—	—
Group II (reference)	47.40 ± 16.53		−16.40%	—
Group III (Q1)	59.00 ± 14.79		4.05%	24.47%
Group IV (Q2)	51.10 ± 17.99		−9.87%	7.80%
Group V (Q3)	48.40 ± 10.99		−14.63%	2.10%
Group VI (Q4)	51.80 ± 14.48		−8.64%	9.28%
Group VII (Q5)	37.40 ± 15.58		−34.03%	−21.09%

Table 7.
The vertical motor activity (VMA) of the groups after 14 days of administration.

	M ± SD	ANOVA	Effect % vs. control	Tukey posttest/ control	Effect % vs. reference	Tukey posttest/ reference
Group I (control)	103.80 ± 14.81	<0.01**	—	—	—	—
Group II (reference)	83.60 ± 14.67		−19.46%	ns	—	—
Group III (Q1)	82.90 ± 12.16		−20.13%	ns	−0.83%	ns
Group IV (Q2)	94.10 ± 14.62		−9.34%	ns	12.55%	ns
Group V (Q3)	104.20 ± 18.61		0.38%	ns	24.64%	ns
Group VI (Q4)	96.20 ± 19.86		−7.32%	ns	15.07%	ns
Group VII (Q5)	107.60 ± 22.46		3.66%	ns	28.70%	<0.05*

**High statistical significance Statistical significance*
***Statistical significance.*

Table 8.
The immobilization time in TST after 14 days of administration.

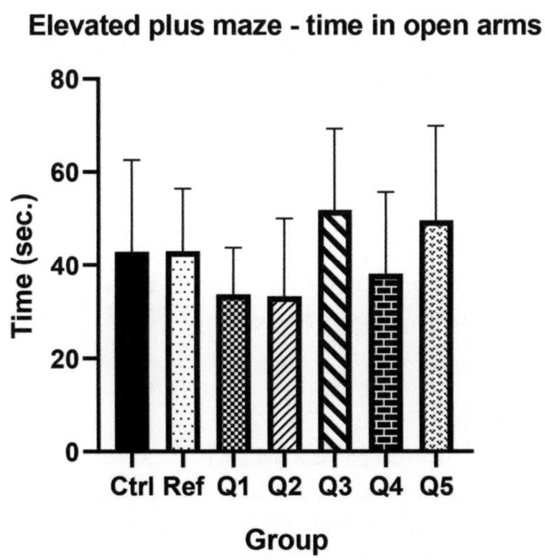

Figure 4.
Time in open arms of EPM after 15 days of administration (the columns represent the mean + SD).

Tail suspension test after 14 days of administration evidenced the compound Q1 which reduced the immobilization time with 20.13% compared with control. The effect is comparable to that of the reference substance. Compound Q5 produced an increase of immobilization time, with 3.66% compared with control and 28.70% compared with amitriptyline (**Table 8**).

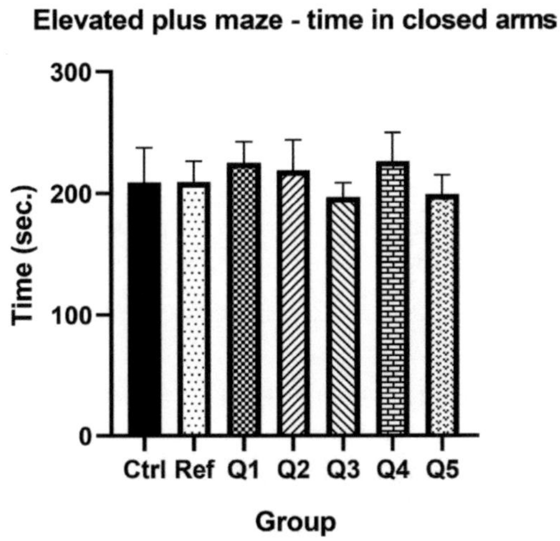

Figure 5.
Time in closed arms of EPM after 15 days of administration (the columns represent the mean + SD).

At the end of the experiment, we wanted to see if the new quinazolinones had any influence on the anxious natural behavior of the mice. It is well known that placed in the plus maze, mice prefer to explore closed and secure arms instead of open ones, associated with imminent danger. The results in elevated plus maze test evidenced a slight anxiolytic effect for two new compounds, Q3 and Q5. They increased the time spent in the open arms of the maze with 20.74 and 15.61% compared with control group. The mean results can be observed in **Figure 4**.

Also, the same two compounds produced a decrease of the time spent in closed arms of the maze, with 5.92 and 4.92% compared with control (**Figure 5**).

The preference of the animals for the open arms instead of closed ones is a sign of lower anxiety, thus we can affirm that compounds Q3 and Q5 have a slight anxiolytic effect. The other compounds did not influence the parameters in EPM.

4. Conclusion

We obtained five new 4(3H)-quinazolinone derivatives through a standardized synthesis process. The compounds were characterized by measuring the melting points, by determining the infrared and NMR spectra, and by elemental analysis.

The pharmacological tests evidenced that the five new quinazolinones have a very low acute toxicity, lethal doses being >2000 mg/kg bw.

Regarding the results obtained in pharmacological tests for evaluation of antidepressant and anxiolytic effects, the compounds had different actions, probably influenced by the presence of different radicals on the nucleus.

Thus, Q1 which have the nitro group in structure manifested the highest antidepressant effect, with a reduction of immobilization time in FST with 20.39% and in TST with 20.13% compared with control.

On the other hand, compounds Q3 and Q5, with two groups methoxy, respective ethoxy, had a slight anxiolytic effect, highlighted by increasing the time spent in open arms, with 20.74 and 15.61% compared with control.

The five new compounds have been shown to have central nervous system activity, and we consider that they deserve further testing in order to detect other effects of interest.

Acknowledgements

We would like to thank Mr. Miron Teodor Căproiu from The Organic Chemistry Center of Romanian Academy "C.D. Nenițescu," for performing the physicochemical determinations.

Conflict of interest

The authors declare no conflict of interest.

Author details

Cornel Chiriță[1*], Carmen Limban[2], Diana Camelia Nuță[2], Emil Ştefănescu[1], Simona Negreş[1], Cristina Elena Zbârcea[1], Cristina Daniela Marineci[1], Oana Cristina Șeremet[1], Mihaela Trandafir (Bratu)[2], Alexandru Vasile Missir[2] and Ileana Cornelia Chiriță[2]

1 Department of Pharmacology and Clinical Pharmacy, Faculty of Pharmacy, University of Medicine and Pharmacy "Carol Davila", Bucharest, Romania

2 Department of Pharmaceutical Chemistry, Faculty of Pharmacy, University of Medicine and Pharmacy "Carol Davila", Bucharest, Romania

*Address all correspondence to: chirita.cornel@gmail.com

References

[1] Rajput R, Mishra AP. A review on biological activity of quinazolinones. International Journal of Pharmacy and Pharmaceutical Sciences. 2012;**4**:66-70

[2] Jain N, Jaiswal J, Pathak A, Singour PK. Synthesis, molecular docking and evaluation of 3-{4-2-amino-4-(substitutedphenyl)-2H-[1,3] oxazin/thiazin-6-yl}2-phenyl-3H-quinazolin-4-one derivatives for their anticonvulsant activity. Central Nervous System Agents in Medicinal Chemistry. 2018;**18**:63-73. DOI: 10.2174/1871524917666170104142033

[3] Tiwary BK, Pradhan K, Nanda AK, Chakraborty A. Implication of quinazoline-4(3H)-ones in medicinal chemistry: A brief review. Journal of Chemical Biology and Therapeutics. 2015;**1**(1). DOI: 10.4172/2572-0406.1000104

[4] Mhaske SB, Argade NP. The chemistry of recently isolated naturally occurring quinazolinone alkaloids. Tetrahedron Letters. 2006;**62**:9787-9826

[5] El-Sayeda NNE, Almaneaia NM, Bachac AB, Al-Obeedd O, Ahmadd R, Abdullad M, et al. Synthesis and evaluation of anticancer, antiphospholipases, antiproteases, and antimetabolic syndrome activities of some 3H-quinazolin-4-one derivatives. Journal of Enzyme Inhibition and Medicinal Chemistry. 2019;**34**(1):672-683. DOI: 10.1080/14756366.2019.1574780

[6] Wolfe JF, Rathman TL, Sleevi MC, Campbell JA, Greenwod TD. Synthesis and anticonvulsant activity of some new 2-substituted 3-aryl-4(3H)-quinazolinones. Journal of Medicinal Chemistry. 1990;**33**(1):161-166

[7] Gatadi S, Lakshmi TV, Nanduri S. 4(3H)-quinazolinone derivatives: Promising antibacterial drug leads. European Journal of Medicinal Chemistry. 2019;**170**:157-172. DOI: 10.1016/j.ejmech.2019.03.018

[8] Poudapally S, Gurram V, Garlapati R, Tulluri C, Addepally U, Vidya K, et al. Cu-free Sonogashira type cross-coupling of 6-halo-2-cyclopropyl-3-(pyridyl-3-ylmethyl) quinazolin-4(3H)-ones as potential antimicrobial agents. Journal of Heterocyclic Chemistry. 2017;**54**:2272-2286

[9] Bouley R, Kumarasiri M, Peng Z, Otero LH, Song W, Suckow MA, et al. Discovery of antibiotic (E)-3-(3-carboxyphenyl)-2-(4-cyanostyryl) quinazolin-4(3 H)-one. Journal of the American Chemical Society. 2015;**137**:1738-1741

[10] Hieu DT, Anh DT, Hai PT, Huong LT, Park EJ, Choi JE, et al. Quinazoline-based hydroxamic acids: Design, synthesis, and evaluation of histone deacetylase inhibitory effects and cytotoxicity. Chemistry & Biodiversity. 2018;**15**(6):e1800027. DOI: 10.1002/cbdv.201800027

[11] Hieu DT, Anh DT, Hai PT, Thuan NT, Huong LT, Park EJ, et al. Quinazolin-4(3H)-one-based hydroxamic acids: Design, synthesis and evaluation of histone deacetylase inhibitory effects and cytotoxicity. Chemistry & Biodiversity. 2019;**16**(4):e1800502. DOI: 10.1002/cbdv.201800502

[12] Ghosh AK, Brindisi M. Organic carbamates in drug design and medicinal chemistry. Journal of Medicinal Chemistry. 2015;**58**:2895-2940. DOI: 10.1021/jm501371s

[13] Bratu M, Nuță DC, Căproiu MT, Missir A, Limban C, Chiriță IC, et al. New acylated derivatives of 2-methyl-4-oxo-quinazolin-3(4H)-il-acetohydroxamic acid. Farmácia. 2014;**62**(4):664-673

[14] Zhu S, Wang J, Chandrashekar G, Smith E, Liu X, Zhang Y. Synthesis and evaluation of 4-quinazolinone compounds as potential antimalarial agents. European Journal of Medicinal Chemistry. 2010;**45**(9):3864-3869

[15] OECD. Guidelines for Testing of Chemicals, 2008

[16] Porsolt RD, Bertin A, Jalfre M. Behavioral despair in mice: A primary screening test for antidepressants. Archives Internationales de Pharmacodynamie et de Thérapie. 1977;**229**:327-336

[17] Papp M, Wieronska J. Antidepressants-like activity of amisulpride in two animal models of depression. Journal of Psychopharmacology. 2000;**14**(1):45-62

[18] Chiriță C, Cristea AN, Militaru M, Negreş S, Zbârcea CE, Nuță CD. Pharmacological evaluation of acute and subacute toxicity and antidepressants effect after acute administration of novel N-substituted benzamides. Farmácia. 2010;**58**(1):21-28

[19] Cryan J, Mombereau C, Vassout A. The tail suspension test as a model for assessing antidepressant activity: Review of pharmacological and genetic studies in mice. Neuroscience and Biobehavioral Reviews. 2005;**29**(4-5):571-625

[20] Pellow S, Chopin P, File SE, Briley M. Validation of open: Closed arms entries in an elevated plus-maze as a measure of anxiety in the rat. Journal of Neuroscience Methods. 1985;**14**:149-167

Chapter 7

Quinazoline Based Synthesis of some Heterocyclic Schiff Bases

Sainath Bhanudas Zangade

Abstract

3-amino-2-methylquinolin-4(3H)-one on condensation with different substituted naphthanones in presence of acetic acid under classical procedure to affords novel series of Schiff bases containing quinazoline moiety. The procedure is simple and easy to obtain the resultant quinazoline Schiff bases in good yields. The product **2a–d** is purified by crystallization in pure 94% ethanol and characterized by thin layer chromatography. The newly synthesized imines (Schiff bases) are confirmed on the basis of spectral techniques, ^{1}H NMR, IR, and mass spectroscopy.

Keywords: synthesis, quinazolines, heterocyclic Schiff bases, naphthnones

1. Introduction

Quinazoline and quinazolinones are the compounds made up of two fused six-membered simple aromatic rings, structure compound containing benzene fused to pyrimidine. This quinazoline ring system consist a structural fragments of about 150–160 natural alkaloids. The first derivative of quinazoline was prepared in 1869 by Griess as 2-cyano-3,4-dihydro-4-oxoquinazoline. This compound bicyclic called as bicyanoamido benzoyl and obtained by the reaction of cyanogens with anthranilic acid [1].

Many of these derivatives possess broad range of biological properties such as antimalarial, antimicrobial, diuretic, anticancer, antiviral, antifungal, antiprotozoal, anti-inflammatory, muscle relaxant, antitubercular, antidepressant, anticonvulsant, weedicide and many others [2–20]. Quinazoline and quinazolinone compounds are also useful nucleus in preparation of various functional materials for synthetic chemistry and also present in various drugs molecules (**Figure 1**).

Schiff bases (imines) are well known for their application and are useful intermediates in organic synthesis. These compounds have intrinsic biological activities including anticancer, antitumour, antitubercular, antioxidant and antiproliferative activity. The combination of quinazoline nucleus with imines further may increase the pharmacological activity, in view of these studies I plan to synthesize four new Schiff bases containing quinazoline (**Figure 2**).

IntechOpen

Afatinib (Tyrosine kinase inhibitor)

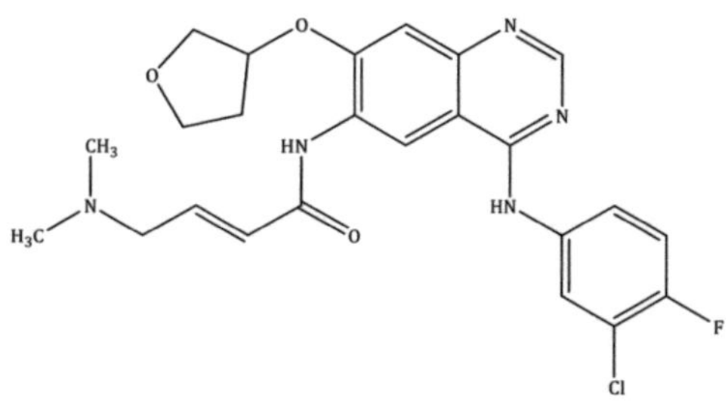

Albaconazole (Antifungal)

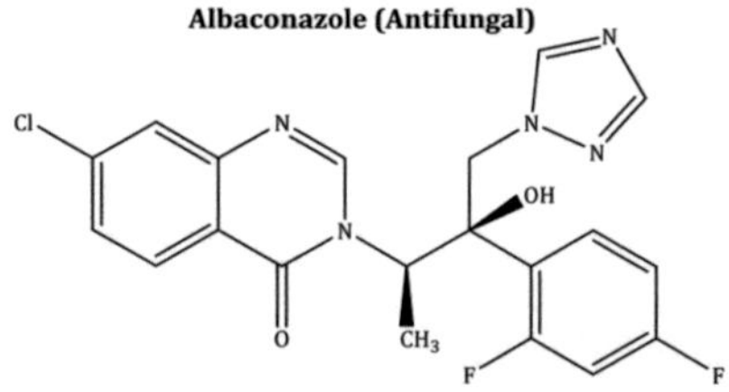

Alfuzocin (Anticancer)

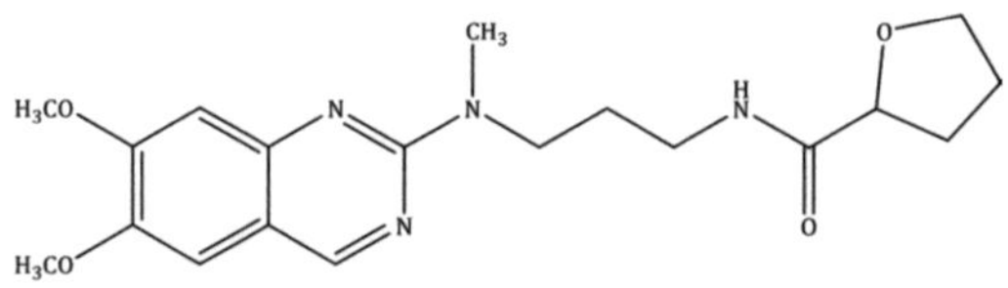

Balaglitazone (Antidiabetic and hypolipidimic)

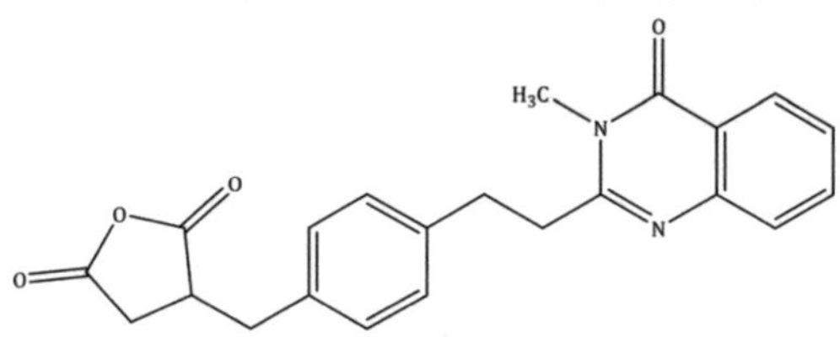

Cediranib (Hematological cancer, liver metastases)

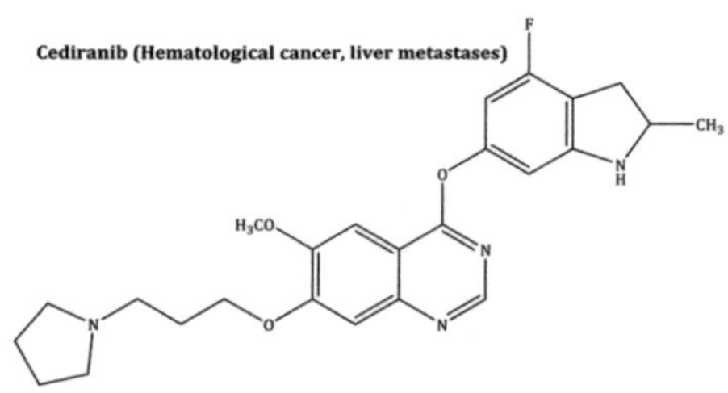

Dacomitinib (Anticancer)

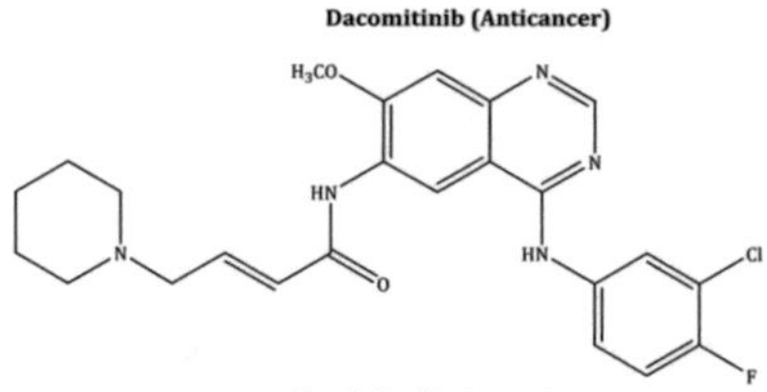

Verubulin (Anticancer)

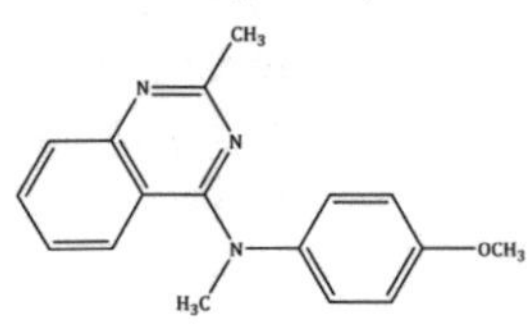

Figure 1.
Biological active drugs containing quinazoline and quinazolinone.

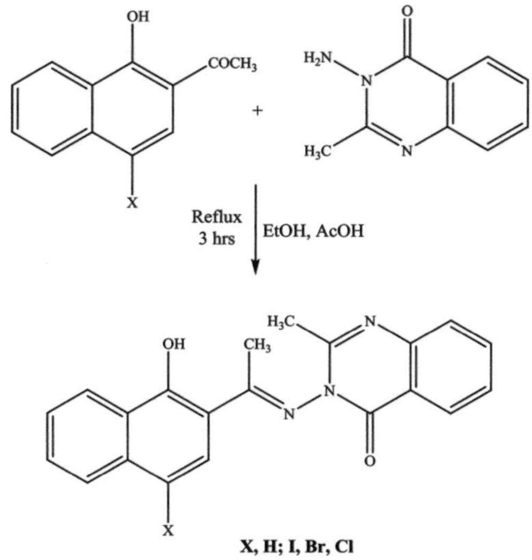

Figure 2.
Synthesis of heterocyclic Schiff bases ***2a–d***.

2. Chemistry

The chemistry of quinazolines was reviewed by a scientist, Williamson in 1957 and then by Lindquist in 1959 and brought up to date by Armarego in 1963. Quinazolines is stable in cold dilute acid and alkaline solutions, but it is destroyed when these solutions are boiled. *O*-Amino benzaldehyde, ammonia, and formic acid are formed when quinazoline is boiled with hydrochloric acid.

2.1 Experimental methods quinazoline based Schiff bases

Melting points were determined in open capillary tubes and are uncorrected. FT-IR spectra were recorded in KBr pellets on a Perkin-Elmer [8201] spectrometer. 1H NMR spectra were recorded on a Gemini 300-MHz instrument in DMSO-d6 as the solvent and TMS was used as an internal standard. The mass spectra were recorded on SHIMADZU (GCMS-QP 1000 EX) GC-EI-MS spectrometer. Elemental analyses were performed on a Perkin-Elmer 240 CHN elemental analyzer. Purification of the compound was indicated using TLC (ethyl acetate/cyclohexane (0.25 mL:0.25 mL, v/v) as the mobile phase).

2.2 Procedure for preparation of quinazoline based Schiff bases

In 50 mL of round bottom flask, mixture of 3-amino-2-methyl-3H-quinazolin-4-one (0.01 M) and substituted acteonaphthanones (0.01 M) was dissolved in ethyl alcohol (15 mL). To this reaction mixture acetic acid (0.001 M) was added and resultant reaction mixture was refluxed for 2–3 h. On completion of reaction as monitored by TLC (ethyl acetate:cyclohexane, 0.25 mL:0.25 mL, v/v as the mobile phase) the reaction mixture was work-up using cold water to obtained crud solid product. The separated solid was filtered and recrystallized from ethanol to yield pure Schiff's bases.

- (E)-3-((1-(1-hydroxy-4-iodonaphthalen-2-yl)ethylidene)amino)-2-ethylquinazolin-4(3H)-one.

Appearance: light yellow solid, FT-IR (KBr, ν, cm^{-1}): 3237, 3065, 2925, 1775, 1628, 1547, 1465, 815, 668. ^1H NMR (300 MHz, DMSO-d_6, δ, ppm): 12.5 (s, 2H, OH), 6.69–8.29 (m, 9H, ArH), 3.26 (s, 3H, CH_3)2.61 (s, 3H, CH_3). (MS, EI) *m/z* (%): 453 (M+, 57%).

- (E)-3-((1-(4-bromo-1-hydroxynaphthalen-2-yl)ethylidene)amino)-2-methylquinazolin-4(3H)-one.

Appearance: light yellow solid, FT-IR (KBr, ν, cm^{-1}): 3242, 3068, 2933, 1772, 1630, 1549, 1460, 818, 663. ^1H NMR (300 MHz, DMSO-d_6, δ, ppm): 12.5 (s, 2H, OH), 6.67–8.29 (m, 9H, ArH), 3.26 (s, 3H, CH_3)2.61 (s, 3H, CH_3). (MS, EI) *m/z* (%): 406 (M+, 69%).

- (E)-3-((1-(4-chloro-1-hydroxynaphthalen-2-yl)ethylidene)amino)-2-methylquinazolin-4(3H)-one.

Appearance: light yellow solid, FT-IR (KBr, ν, cm^{-1}): 3247, 3063, 2935, 1774, 1632, 1544, 1462, 816, 665. ^1H NMR (300 MHz, DMSO-d_6, δ, ppm): 12.5 (s, 2H, OH), 6.68–8.29 (m, 9H, ArH), 3.27 (s, 3H, CH_3)2.61 (s, 3H, CH_3). (MS, EI) *m/z* (%): 361.5 (M+, 65%).

- (E)-3-((1-(1-hydroxynaphthalen-2-yl)ethylidene)amino)-2-methylquinazolin-4(3H)-one.

Appearance: light yellow solid, FT-IR (KBr, ν, cm^{-1}): 3242, 3067, 2937, 1773, 1635, 1546, 1465, 818, 667. ^1H NMR (300 MHz, DMSO-d_6, δ, ppm): 12.5 (s, 2H, OH), 6.68–8.29 (m, 10H, ArH), 3.25 (s, 3H, CH_3)2.62 (s, 3H, CH_3). (MS, EI) *m/z* (%): 327(M+, 64%).

3. Conclusion

The author has reported the synthesis of novel series of quinazolines based Schiff bases from hydroxynaphthanones and 3-amino-2-methyl-3H-quinazolin-4-one using classical procedure under slightly acidic conditions.

4. Result and discussion

Quinazoline derivatives are further classified into the following main five categories. This classification is on the basis of their substitution patterns present in the ring system [3]. These are 2-substituted-4(3*H*)-quinazolinones, 3-substituted-4(3*H*)-quinazolinones, 4-substituted-quinazolines, 2,3-disubstituted-4(3*H*)-quinazolinones, and 2,4-disubstituted-4(3*H*)-quinazolinones. In view of the importance of quinazoline derivatives, the present study describes the synthesis of some important quinazoline based imines. The mixture of corresponding 3-amino-2-methyl-3*H*-quinazolin-4-one and substituted naphthnones treated in presence slightly acidic condition under reflux for 3 h to give formation imines (**Figure 2**). The formation of imines confirmed on the basis of spectral data, the IR spectra reveals the stretching band around 1630 cm^{-1} is due to imines C=N. The NMR reveals the absence of –NH_2 protons around δ 5.2 confirms the condensation reaction carried successfully. The mass spectrum of compounds **2a–d** gives molecular ion which is corresponding to molecular weight of products.

Conflict of interest

The authors confirm that this article content has no conflict of interest.

Author details

Sainath Bhanudas Zangade
Department of Chemistry, Madhavrao Patil ACS College, Parbhani, Maharashtra, India

*Address all correspondence to: drsbz@rediffmail.com

References

[1] Armarego WLF. A text book of quinazolines. Advances in Heterocyclic Chemistry. 1963;**1**:253-309. DOI: 10.1016/S0065-2725(08)60527-9

[2] Rajveer CH, Swarnalatha CH, Stephen RB, Sudhrshini S. Synthesis of 6-bromo-oxo quinazoline derivatives and their pharamcological activities. International Journal of Chemistry Research. 2010;**1**:21-24

[3] Alafeefy AM, Kadi AA, Al-Deeb OA. Synthesis, analgesic and anti-inflammatory evaluation of some novel quinazoline derivatives. European Journal of Medicinal Chemistry. 2010;**45**:4947-4952

[4] Salunkhe PS, Patel HM, Shimpi RD, Lalwani NN. Study of analgesic & anti-inflammatory evaluation of some 2,3-dihydroquinozoline-4-one derivatives. International Journal of Pharmaceutical Research. 2010;**2**:974-987

[5] Aanandhi MV, Velmurugan V, Shanmugapriya S. Synthesis, characterization and biological evaluation of 3,4-dihydro quinazoline2(H)-one derivatives. Journal of Chemical and Pharmaceutical Research. 2011;**2**:676-683

[6] Patel HU, Patel RS, Patel CN. Synthesis and antihypertensive activity of some quinazoline derivatives. Journal of Applied Pharmaceutical Science. 2013;**3**:171-174

[7] Honkane E, Kairisalo P, Nore P, Karppanen H. Synthesis and antihypertensive activity of some new quinazoline derivatives. Journal of Medicinal Chemistry. 1983;**10**:1433-1438

[8] Aly AA. Synthesis and antimicrobial activity of some annelated quinazoline derivatives. Journal of the Chinese Chemical Society. 2007;**54**:437-446

[9] Selvam TP, Kumar PV, Rajaram C. Synthesis and antimicrobial activity of novel thiazoloquinazoline derivatives. International Journal of Pharma Sciences. 2010;**2**:119-122

[10] Break LM, Mosselhi MA. Synthesis, structure and antimicrobial activity of new 3 and 2-arylmethyl and arylacyl-3H[1,2,4]triazino[3,2-b]-quinazoline-2,6(1H) diones as expected as DNA fluorphores. Journal of Chemical Sciences. 2012;**5**:23-28

[11] Kiruthiga B, Ilango K, Valentina P. Synthesis of some new 2-substituted quinazolin-4-one derivatives and their biological activities. International Journal of PharmTech Research. 2009;**4**: 1503-1506

[12] El-helby AG, Abdel Wahab MH. Design and synthesis of some new derivatives of 3H-quinazolin-4-one with promising anticonvulsant activity. Acta Pharmaceutica. 2003;**53**:127-138

[13] Ilangovan P, Ganguly S, Pandi V, Stables JP. Design and synthesis of novel quinazolinone derivatives as broad spectrum anticonvulsants. Scholars Research Library. 2010;**1**:13-21

[14] Nerkar AG, Saxena AK, Ghone SA. In silico screening, synthesis and in vitro evaluation of some quinazolinone and pyridine derivatives as dihydrofolate reductase inhibitors for anticancer activity. Journal of Chemotherapy. 2009;**6**:97-102

[15] Manasa K, Sidhaye RV, Radhika G, Nalini CN. Synthesis, antioxidant and anticancer activity of quinazoline derivatives. Current Pharma Research. 2011;**2**:101-105

[16] Guan J, Zhang Q, O'Neil M. Antimalarial activities of new pyrrolo[3,2-f]quinazoline-1,3-diamine

derivatives. American Society for Microbiology. 2005;**12**:4928-4933

[17] Wang HJ, Wei CX, Deng XQ. Synthesis and evaluation on anticonvulsant and antidepressant activities of 5-alkoxy tetrazolo[1,5-a] quinazolines. Archiv der Pharmazie. 2009;**11**:671-675

[18] El-enany MM, El-meligie SM, Abdou N, El-nassan B. Synthesis and antimicrobial activity of some 3,5-diaryl-4,5-dihydropyrazole derivatives. Oriental Journal of Chemistry. 2010;**4**:1265-1270

[19] Khalil OM, Refaat HM. Synthesis and antiinflammatory activity of some 3,5-diaryl-4,5-dihydropyrazole derivatives. Oriental Journal of Chemistry. 2011;**4**:1581-1591

[20] Kumar A, Verma RS, Jaju BP, Sinha JN. Synthesis and biological properties of 4-(3H)-quinazolone derivatives. Journal of the Indian Chemical Society. 1990;**67**:920-921